Orchestrating Enterprise Risk

Why the Greatest Risk Lives Between Systems, Processes, and Controls

E.H. Schmidt

Sentralis Press

Published by Sentralis Press.

Formatted with Vellum

A Note on Intellectual Lineage

The four planes of assurance in this book did not emerge from nothing. They owe a clear debt to SABSA (https://sabsa.org/), the business-driven security architecture method created by John Sherwood and described in Enterprise Security Architecture: A Business-Driven Approach (Sherwood, Clark, and Lynas). SABSA's core conviction — that every control must trace back to business intent, through explicit layers, with authority and evidence made visible rather than assumed — runs underneath everything that follows. SABSA is a registered trademark of The SABSA Institute. This book is an independent work; it is not affiliated with, endorsed by, or certified by The SABSA Institute, and the four-plane model is the author's own synthesis, not a reproduction of the SABSA framework. See Sherwood, J., Clark, A., and Lynas, D., Enterprise Security Architecture: A Business-Driven Approach (CMP Books / CRC Press, 2005; ISBN 978-1-57820-318-5); and The SABSA Institute, White Paper W100.)

Chapter 1
The Gap Nobody Named

THE SYSTEM RAN ALL NIGHT

Right now, while you're reading this sentence, your enterprise systems are making decisions. They're approving transactions, paying vendors, moving inventory, posting journal entries, and updating financial records. Thousands of decisions are happening without human intervention. And that's by design.

Most of those decisions are correct. The question isn't whether your systems are working. The question is whether you actually know they're still under control.

And yet.

If I asked you whether your business was truly under control—not just one system or one process, but the enterprise as a whole—could you answer with confidence?

With evidence, not just documentation.

In nearly thirty years working with senior risk and finance leaders, I've found most can't. They can tell me the controls that exist. They can show me the reports. What they can't show me is whether those controls are working together in a way that reflects how the business operates today.

That's the problem this book is about.

The problem isn't that organizations lack controls. They lack a unified view of how those controls interact across systems, processes, and business operations.

The greatest risks rarely exist within a single tool. They emerge in the space between them.

EVERYBODY SOLVES A PIECE

The enterprise risk industry has done a remarkable job building excellent tools. GRC platforms catalog your risks and map your controls. Identity and access management systems govern who can do what. SAP security and SoD tools watch for dangerous permission combinations. Continuous controls monitoring catches transaction anomalies. Audit management platforms track findings through remediation. Process mining shows you where your workflows actually go versus where you thought they went.

Each one does its job well. That's genuinely not the problem.

The problem is that none of them talk to each other in any meaningful way. Each tool produces its own outputs, in its own format, against its own definitions. A control flagged in your GRC platform might not match the violation your access governance tool just detected. The transaction your monitoring tool flagged last Tuesday doesn't appear in the audit finding from last month. The segregation issue your security team knows about hasn't made it to the CFO's dashboard.

The risk that matters most lives in those gaps.

I've sat in enough executive working sessions and audit committee meetings to know what happens next. Someone synthesizes the outputs manually, usually a stressed analyst working from a spreadsheet, which then gets massaged through a higher-level review. The picture they produce is already weeks old by the time it reaches the table. And the system has kept running.

There's a phrase I keep coming back to: *compliant and exposed.*

You can check every box your auditors put in front of you and still be operationally vulnerable. You can produce a clean opinion and still have a material failure in progress. The documentation is solid, the controls exist on paper, and the gap is real.

That fragmentation isn't a vendor problem or a budget problem. It's a discipline problem. There's no named field, no established practice, no set of frameworks built to cross-reference these tools and blend their outputs into a single coherent picture of whether the enterprise is actually in control.

That's the gap nobody named. Until now.

WHY THIS BOOK, WHY NOW

I've spent nearly thirty years working across the full span of this problem: Big Four advisory, SAP implementation and security, GRC consulting, and product development. I've watched enterprises spend serious money on excellent point solutions and still show up to audits scrambling to reconcile outputs that should never have been separated.

The timing matters. Two forces are accelerating the problem.

The first is automation density. More of your business runs at machine speed. SAP doesn't wait for a human to review a payment run. Approvals, postings, settlements, and journal entries happen continuously, and the controls designed to govern them were largely built for a slower world. When something goes wrong, the damage is often irreversible before any of the point-solution tools have surfaced the signal.

The second is AI. Enterprises are beginning to deploy AI agents that don't just report or flag; they take action. They approve, route, and execute. Most organizations have no assurance layer sitting between those agents and the transactions they're touching. The fragmentation problem that already existed is about to get a very fast engine attached to it.

This isn't meant to alarm you. The leaders who understand

what's happening here have a real advantage. An enterprise with a coherent, real-time picture of its control environment will consistently outperform one still relying on annual audits and manual reconciliation. The question is who builds that capability first.

HOW TO USE THIS BOOK

This book teaches a discipline I'm calling *Enterprise Risk Orchestration.* The core idea is straightforward: you don't need to replace the tools you already have. You need to connect their outputs, normalize their language, and build composite views that show you what no single tool can see alone.

The first three chapters name and prove the problem. If you're already living with it, they'll feel familiar, maybe uncomfortably so. The middle chapters introduce the framework: what I call the *four planes of assurance*, applied to the landscape you're already running. The final chapters address where this is heading, what continuous assurance actually means in practice, and what boards and CFOs should be asking right now that most aren't.

You can read this front to back or treat it as a reference. If you're a CRO or CFO trying to frame the problem for your board, start with "The Gap Nobody Named" and "What the Audit Report Doesn't Say." If you're a Chief Audit Executive trying to modernize your function, the chapter on seeing the whole picture is where the framework comes together. If you're an ERP risk or controls leader who already knows the fragmentation problem and wants language to take it upstairs, the whole book is for you.

A note on examples. When this book gets specific, it leans most often on SAP. That's deliberate. SAP runs the back office of more of the world's largest enterprises than any other system, and it carries the deepest stack of governance, risk, and compliance tooling of any ERP — which makes it the hardest possible test of the argument here. If the gaps still open up in the most heavily governed environment on the market, they open up everywhere else too. But the discipline in this book is not about SAP. It applies to any enterprise landscape — NetSuite, Oracle, Workday, or a patchwork of all three. SAP simply gives the clearest illustration.

I've tried to write something useful at the practitioner level and readable at the board level. Whether I pulled that off is something you'll decide by the end.

Chapter 2
Compliant and Exposed

The letter arrives and the CFO exhales. External auditors have issued their opinion. Internal controls over financial reporting are effective. No material weaknesses. The audit committee nods. The board checks the box. Everyone goes back to running the business.

That exhale is the most expensive moment in enterprise risk management.

I've watched it happen across dozens of organizations over nearly thirty years. The relief is genuine. The problem is that it's misplaced. A clean audit opinion tells you something real and useful. It just doesn't tell you what most people in that room think it tells them.

THE CLEAN OPINION

Let's be precise about what a clean opinion actually says.

When auditors attest that internal controls over financial reporting are effective, they are saying that the controls they tested, over the period they examined, using the sample sizes they selected, produced results that met the thresholds defined by the standards they applied.

Read that again slowly. Tested. Period. Sample. Thresholds. Standards.

Every one of those words is a boundary condition. The auditor is not saying your controls work everywhere, all the time, across every transaction, in every corner of your system. They're saying what they found, in the scope they covered, holds up to a specific definition of effective.

That's a meaningful finding. It's genuinely worth having. SOX 404 has improved financial reporting discipline enormously since 2004, and the data bears that out. But it's a bounded finding, and the boundary matters more than most executives realize.

According to a nearly two-decade review of SOX 404 disclosures, roughly 24% of all companies filing ICFR assessments reported adverse findings in FY2022. That's about one in four. And for companies filing for the first time, the adverse rate was 38%. First-time auditor attestations specifically came in at 28% adverse. These are companies that had controls in place, had processes documented, had people accountable. They still didn't pass.

But here's what concerns me more than the companies that failed the audit. It's the ones that passed.

A clean opinion is issued to roughly three out of every four filing companies in any given year. Some of those companies are genuinely well-controlled. Others passed the audit and are still sitting on significant operational exposure they don't know about. The audit didn't find it. That's not the auditor's fault. It's the nature of what an audit is designed to do.

The problem isn't that audits are bad. The problem is that organizations confuse passing the audit with being in control.

WHAT 'PASSING' ACTUALLY MEASURES

Auditors sample. That's the fundamental reality of how large-scale assurance works. From a population of millions of transactions, they test twenty-five or fifty items. They pick a period, usually a quarter or a full year. They look at what the controls documentation says should happen, then test whether it happened.

That's a reasonable methodology given the practical constraints of what humans can do manually in a reasonable timeframe. It's the best available approach when you're working with paper evidence, interviews, and spreadsheets.

What it can't do is observe the system continuously. What it can't do is see what happens between tests. What it can't do is detect drift in the configuration that occurred after the audit fieldwork closed, or the access that was provisioned outside the formal request process, or the approval that technically satisfied the control but was routed around its original intent.

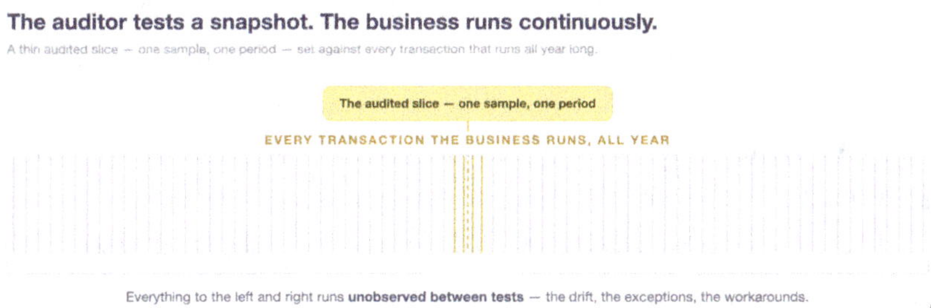

I've seen the gap between those two realities show up in predictable ways. An auditor reviews journal entry approvals and finds the sample clean. What the sample doesn't surface is that a handful of entries each month are posted by someone with both the ability to create and the ability to approve, because a system role was misconfigured eighteen months ago and no one connected the two data sets. The GRC tool shows the role as compliant. The access management tool shows the user as provisioned. Neither tool is wrong. Neither tool sees the whole picture.

That's not a failure of the auditor or the tools. It's a structural gap in how we've built assurance. Each piece does its job. Nobody owns the space between the pieces.

The most common categories you see in material weakness disclosures confirm this pattern. Access controls. Segregation of duties. Journal entries. Reconciliations. Change management. IT general controls. These show up year after year in adverse ICFR filings. In FY2022, IT and security issues were cited in more than half of all adverse auditor attestations. SoD issues led management-only reports. The same categories. Year after year.

If these were simple problems, they'd be solved by now. Organizations have spent decades and billions of dollars on tools specifically designed to address each one. GRC platforms manage policy and control documentation. Identity and access management systems govern who has what. Security platforms monitor suspicious activity. Change management tools track configuration and code changes.

And yet the same categories keep showing up in adverse disclosures. Also, many of these organizations still can't identify a solid set of SoD rule sets based on their own processes.

The reason is that the categories aren't the problem. They're the symptoms.

THE GAP BETWEEN THE REPORT AND REALITY

The language auditors use is precise by design. It has to be. It goes into a public filing, and every word carries legal weight. So the language is accurate as a description of what failed. It's rarely accurate as a description of why.

"The company failed to maintain effective controls over inventory movements."

What that typically means in practice: receiving recorded the goods in the warehouse system, but the inventory valuation in the ERP ran on a different posting cycle, and for weeks the two never agreed on what was actually on hand. Each system was internally consistent. The control tested each one and found it clean. Nothing tested whether the two told the same story. The system kept running.

"The company failed to maintain effective controls over revenue recognition."

What that often means: sales, logistics, billing, and finance were each operating from a different version of the same transaction. One system said the order shipped. Another said the invoice wasn't generated. A third said the revenue was booked. A fourth said the customer hadn't been billed yet. Four systems. Four perspectives on the same event. No single view of what was true.

"The company failed to maintain effective controls over user access."

What that frequently means: someone did the annual access review, checked all the boxes, certified all the roles as appropriate. Then three weeks later a reorganization happened, roles changed, and a user ended up with access they shouldn't have had. The review was done correctly. The world changed after the review. Nobody was watching in between.

In each case, the auditor's language is accurate. The control failed. But the reason it failed isn't captured in those words. The actual failure happened in the gap between systems, between processes, between the moment a human last reviewed something and the moment the system acted on its own.

I've spent much of my career questioning that gap. It's where real exposure lives, and most consulting pushes security access as a primary control, yet it is only a small portion of the solution.

WHAT 'PASSING' DOESN'T PROTECT

Here's where I want to push back on something the industry has quietly accepted as normal.

The standard narrative goes like this: you invest in controls, you engage auditors, you pass the audit, you're good. If you keep passing the audit, your control environment is improving. If you have a finding, you remediate it, pass again, and the problem is solved.

This narrative is wrong in a specific and important way. Passing the audit and being in control are not the same thing.

A company can pass every audit for five consecutive years and still have a payment fraud scheme operating quietly in the accounts payable process, because the fraud exploits the gap between the vendor master data system and the payment approval workflow. Neither system flags it. The auditor's sample didn't capture it. The GRC tool shows green.

A company can have an IT general controls program that any auditor would find acceptable while running on a system configuration that was last formally reviewed before a major platform upgrade. The configuration drift happened. The upgrade documentation is in a separate tool. Nobody cross-referenced them.

A company can have clean SoD reports coming out of its GRC tool, while a small number of users have roles in the ERP that the GRC tool's rule set doesn't flag, because the rule was written for a previous version of the business process and nobody updated it.

These aren't hypothetical scenarios constructed to make a point. They're patterns I've seen across organizations in multiple industries, at multiple scales, across every major ERP platform. The details differ. The underlying structure is the same.

Control exists. The controls are documented. The controls pass review. The exposure continues.

THE REAL COST OF THE GAP

The financial impact of this dynamic doesn't show up cleanly in a single line item. It accumulates.

It accumulates in remediation costs when a finding finally surfaces, often after it's been running for a year or more. It accumulates in the labor cost of annual audit preparation, where finance and IT teams spend weeks pulling evidence that should be available in minutes. It accumulates in the cost of restatements when something that should have been caught turns out to have been affecting reported numbers for multiple periods.

The operational cost is real too. When your assurance model depends on periodic manual testing, you get periodic manual certainty. Between those periods, you're running on assumption. Decisions get made on the assumption that controls are working because they worked last quarter. Transactions get approved on the assumption that access profiles are correct because they were certified last month. Configurations are trusted because they passed the last change management review.

Assumptions are fine when the stakes are low. They're expensive when the stakes are high.

One consistent pattern across organizations that have done this analysis: the cost of maintaining the illusion of control often exceeds the cost of having real control. The audit preparation scramble alone. The fire drills when a finding surfaces close to filing deadlines. The consulting fees to help remediate something that the organization's own data, properly connected, would have flagged months earlier.

It's worth noting that in FY2022, the rate of adverse auditor attestations hit 7.6%, the highest it had been since 2007. IT and security issues were cited in 54.5% of those adverse attestations. That's not a bad year for one or two unlucky companies. That's a systemic problem expressing itself across hundreds of organizations simultaneously.

The finding categories haven't changed much in nineteen years. The tools for addressing them have gotten dramatically better. The gap persists.

The reasonable conclusion is that the problem isn't the tools. It's the architecture of how the tools relate to each other and to the risk picture leadership needs to see.

THE PHRASE THAT CHANGES THE CONVERSATION

There's a phrase I've started using when I'm working with leadership teams, and I've watched it land differently than almost anything else I bring into those conversations.

Compliant and exposed.

I say it as a single concept, not two separate conditions. Because that's what it is. An organization can be genuinely compliant, by every reasonable definition of that word, and simultaneously exposed to risks that the compliance process wasn't designed to detect.

This is not an argument against compliance. Compliance is a floor. A necessary floor. You need to meet it. SOX 404 has made financial reporting more reliable. GRC programs have professionalized risk management in ways that matter. None of that goes away.

The argument is that compliance is a floor, and too many organizations treat it as a ceiling.

When the primary question your organization is asking is "did we pass the audit," you're optimizing for a measurement that captures a subset of your actual exposure. When the primary question becomes "are we actually in control," the answer requires a different kind of evidence from a different kind of process.

The distinction sounds subtle. The operational difference is substantial.

"Did we pass the audit" is answered by checking what auditors tested against what they found.

"Are we actually in control" is answered by observing whether the system is behaving as intended, continuously, across every material process, right now.

The first question looks backward. The second looks at the current state. The first is satisfied by documentation. The second requires evidence. The first question can be answered by a small team doing intensive periodic work. The second requires continuous observation built into how the organization runs.

Most organizations are set up to answer the first question very well. Almost none are set up to answer the second.

WHY THIS STAYS HIDDEN

I want to be honest about why this problem persists, because it's not for lack of smart people or good intentions.

The people who run audit functions at major enterprises are skilled professionals doing difficult work under real constraints. The people who manage GRC programs understand their tools and work hard to keep them current. The people who configure SAP access controls and manage identity systems are doing technically demanding work that matters.

The problem isn't competence. It's that nobody owns the connection points.

An internal audit function owns its methodology and its findings. An IT security team owns its monitoring and its incidents. A GRC team owns its framework mappings and control documentation. An SAP access controls team owns its role designs and its conflict rules.

Each team looks at its piece of the picture. Nobody's job description says "cross-reference all of these outputs and tell me what they add up to."

So they don't.

The audit team produces its report. The security team produces its report. The GRC tool generates its dashboards. The access review produces its certifications. All of that information exists. None of it is connected.

I've sat in rooms where eight different reports were on the table, each produced by a different team using a different tool, each showing green or showing findings within an acceptable range. And the room was still sitting on a real exposure, because the exposure was visible only in the intersection of those eight data sets, which nobody had combined.

That's not a people problem. It's an architecture problem.

I've also seen the dynamic that makes this hard to fix quickly: organizations that try to solve it by buying another tool. Another dashboard. Another platform that promises to "unify" the risk picture by being yet another system that other systems need to integrate with.

That's adding a ninth report. It doesn't solve the problem.

The fix isn't a tool. It's a discipline. A way of thinking about what the existing tools are actually producing, where the gaps between their outputs are, and how to read those gaps as the risk signal they actually are.

That discipline is what this book is about. And its foundation is the ground we covered earlier: understanding clearly and specifically what compliance measures, what it doesn't measure, and what it will cost you to confuse the two.

THREE THINGS TO DO BEFORE THE NEXT CHAPTER

Before moving forward, consider where your own organization sits on this question. These aren't exercises with right answers. They're questions designed to help you see the picture more clearly.

First, look at how leadership frames the assurance conversation. Is the primary question being asked "did we pass" or "are we in control"? The framing shapes everything that follows, including what evidence gets produced, who's accountable for what, and how quickly problems surface. If the room celebrates the clean opinion and moves on, that's a framing worth examining.

Second, identify where your organization's siloed outputs live. Your GRC tool produces reports. Your identity management system produces access certifications. Your SAP security team produces SoD analysis. Your internal audit team produces findings. Where, if anywhere, do those outputs get cross-referenced? Who, if anyone, is looking at the intersection? If the answer is "nobody formally," that's not a failure. It's a gap waiting to be addressed.

Third, try to name one risk that lives between two systems. One process where the ERP records one thing and another system records something adjacent, and no one regularly compares the two. Every organization I've worked with has these. They're usually not hard to find once you start looking. The harder question is: how long has that gap existed, and what's been running through it unobserved?

The goal at this stage isn't to fix anything. It's to see clearly. The rest of the book builds on that clarity.

Compliant and exposed is the starting condition for most large organizations today. The path out of it starts with being honest about what compliance actually measures and what it doesn't. That honesty is harder than it sounds in organizations where passing the audit has been the definition of success for twenty years.

The good news: the data you need to get honest already exists inside your organization.

That's the entire argument, in miniature. The chapters that follow build the framework for how to address it.

Chapter 3
What the Audit Report Doesn't Say

Every audit report I've ever read follows roughly the same structure. There's a scope statement, a description of the testing approach, and then the findings. The findings are written in a specific kind of language. Careful language. Precise, in its own way. But reading one as a practitioner feels a little like reading a weather report written two weeks after the storm passed.

The language tells you a control failed. It rarely tells you why. It tells you what category the failure fell into. It almost never tells you what actually happened in the business.

That gap between the report and reality is what this chapter is about.

THE SIX CATEGORIES AUDITORS ALWAYS FIND

When an audit goes sideways, the findings almost always land in one of six buckets:

- Access Controls
- Segregation of Duties (SoD)
- Journal Entries

- Reconciliations
- Change Management
- IT General Controls (ITGCs)

This isn't a coincidence. These six categories exist because audit methodology was designed to look for them. They're baked into SOX frameworks, internal audit programs, external auditor checklists, and every GRC implementation I've seen across nearly 30 years. If you want to find problems, you look in the places where problems are expected.

The thing is, these categories do surface real problems. Access controls fail. SoD conflicts exist. Journal entries get made outside normal workflows. Reconciliations break down. Change management goes unsupervised. ITGC gaps open the door to everything else.

But here's what I've noticed: when you look at the same six categories across different companies, different industries, different ERP platforms, different audit firms, you start to see a pattern that goes beyond the categories themselves. The pattern is that these findings keep recurring. Companies remediate them, pass the next audit, and then they show up again. Different form, same category.

That tells you something. When the same types of findings reappear year after year at company after company, it's worth asking whether the categories are identifying root causes or just symptoms.

In my experience, they're mostly symptoms.

WHAT THE FINDINGS ACTUALLY LOOK LIKE

The SOX 404 data backs this up. In a 19-year review of adverse ICFR disclosures, the number one internal controls issue cited in adverse auditor attestations for FY2022 was information technology, software, and security issues. It was the first time in five years that IT topped the list. The second most common was accounting personnel resources. Both were cited in more than half of all adverse attestations that year.

Revenue recognition issues showed up as the most common accounting-level finding, appearing in 23.3% of adverse auditor attestations in FY2022 and holding the top spot for five consecutive years. Accounts receivable issues came in at 12.1%. Inventory and long-term assets at 11.7%.

These are useful numbers. They confirm what practitioners already know: the same problems keep surfacing, in the same categories, with remarkable consistency. But the numbers also hide something. They tell you where auditors found issues. They don't tell you where the issues actually started.

AUDITOR LANGUAGE VS. WHAT ACTUALLY HAPPENED

Audit reports are written in a specific vocabulary. That vocabulary serves a real purpose. It's precise, defensible, and structured to communicate findings to boards, audit committees, and regulators. But the precision of audit language can also create distance from the actual operational event.

Here's what I mean.

The audit report says: "The company failed to maintain effective controls over inventory movements."

What actually happened: the warehouse was shipping product before the ERP transaction was recorded. The physical movement and the system record were consistently out of sync. Nobody planned for it to be that way. The warehouse team was under pressure to hit shipping targets. The ERP input happened later, when someone got to it. Over time, the gap became normal.

The audit report says: "The company failed to maintain effective controls over revenue recognition."

What actually happened: sales, logistics, billing, and finance were all working from different versions of the same information. Sales booked the deal in one system. Logistics tracked fulfillment in another. Billing processed on its own schedule. Finance closed the books from reports that pulled from all three. When those reports disagreed, people made judgment calls. Some of those calls moved revenue across periods. The root cause wasn't a bad actor or a broken control. It was four systems that never actually talked to each other, operating in parallel, each one individually correct, collectively incoherent.

There's a big difference between those two descriptions and what ends up in the formal report. The formal finding tells you the category. The operational reality tells you the cause. Remediation that addresses only the category will leave the cause intact.

What the report says vs. what actually happened

The same finding, in two languages: the formal wording on the left, the operational reality on the right.

	WHAT THE REPORT SAYS		WHAT ACTUALLY HAPPENED
INVENTORY	"A deficiency was identified in controls over inventory valuation."	→	A costing variance ran for two quarters because no one reconciled the new plant's feed.
REVENUE	"Controls over revenue recognition did not operate effectively during the period."	→	Reps changed pricing after order entry; the override was never given a second review.
ACCESS	"Segregation-of-duties conflicts were not consistently remediated."	→	One clerk could both create a vendor and release its payment — for fourteen months.

This is one of the most expensive mistakes I see organizations make. They respond to the audit finding with a control fix targeted at the symptom. The SoD conflict gets resolved in the access control tool. The reconciliation process gets tightened. The journal entry policy gets updated. Six months later, a variation of the same problem surfaces somewhere else, usually in a slightly different form.

The fix was applied to the finding, not to the system that produced it.

WHY THE CATEGORIES KEEP RECURRING

The six categories aren't random. They keep showing up because they map to structural weaknesses in how large enterprises are organized and operated.

Access controls fail repeatedly because managing access across a complex enterprise with multiple systems, multiple roles, frequent personnel changes, and inconsistent provisioning processes is genuinely hard. No single tool sees the full picture. You might have SailPoint managing identity, SAP GRC managing ERP access, an Azure AD layer governing cloud access, and a half-dozen other systems each enforcing their own rules. Each one does its job. None of them automatically talk to the others. So a person can have appropriate access in four systems and dangerous access in a fifth, and no single tool flags the combination.

SoD conflicts persist for the same reason. SoD analysis is typically run within a single system, against a predefined conflict ruleset. But business processes don't stay inside one system. A purchase-to-pay process runs through SAP. It also touches procurement workflow tools, vendor management systems, payment gateways, and banking platforms. An SoD analysis that only looks at SAP gives you a partial picture. The conflict that actually matters might require access in two different systems that no single audit ever examines together.

Journal entries are a perennial finding because manual journal entries represent exactly the kind of discretionary human action that can bypass automated controls. The automation works, so people route around it. The harder question isn't whether manual journal entries exist. It's whether the pattern of those entries, the timing, the amounts, the frequency, the approvers, and the downstream effects, tells you something about how the business actually runs versus how the controls were designed.

Reconciliations break down because reconciling across systems is labor-intensive and deeply dependent on people doing the right thing at the right time. When those people are stretched, or when the systems produce outputs in different formats, or when the data extraction takes too long, reconciliations slip. And when reconciliations slip, nobody sees the divergence until it's already large.

Change management failures persist because ERP and application changes are frequent, urgent, and politically complex. A production support team under pressure to fix a system issue doesn't always have time for a full change control cycle. An emergency transport in SAP happens outside the normal approval path. The change gets made, the system runs, and the audit trail is incomplete. Nobody maliciously bypassed the process. The process just wasn't fast enough for the situation.

ITGCs are the structural foundation under all of this. When ITGC controls are weak, everything built on top of them is suspect. Access controls, change management, availability, processing integrity: all of these depend on ITGC discipline. When ITGCs appear in an adverse finding, it's usually a signal that the foundational controls infrastructure is underinvested or inadequately maintained.

What all six categories have in common is this: they reflect the normal failure modes of complex, multi-system, multi-process enterprises. They're the places where fragmented tools try to enforce coherent control over processes that span all of them, and fall short.

SYMPTOMS, NOT ROOT CAUSES

I've spent time helping companies respond to audit findings. The remediation work is real work. Fixing an SoD conflict requires understanding the role design, the business process, and the access model in the affected system. Tightening a reconciliation process requires documenting the control, assigning ownership, setting a cadence, and making sure the right people actually do the work.

None of that is trivial. But I've also watched many of those fixes hold for one or two audit cycles and then erode. The SoD conflict gets remediated, and then a year later there's a new one in a system the previous remediation didn't touch. The reconciliation gets tightened, and then a new product line gets stood up with a different billing system that wasn't included in the original reconciliation scope.

The pattern keeps repeating because the remediation keeps targeting the finding, not the underlying condition. Next the company needs to spend potentially millions on consultants to redesign or reset their security model.

The underlying condition is fragmentation. The six categories are where fragmentation shows up. They're the places where the gaps between systems, processes, and people become visible enough to catch an auditor's attention. But the source of the finding is usually cross-functional, cross-system, or both.

Think about the inventory example again. The warehouse shipped before the ERP transaction was recorded. That might surface as an access control finding, possibly a reconciliation finding, maybe an ITGC finding depending on how the auditor frames it. But the root cause is an operational process that never aligned with the system design. The people involved were doing their jobs. The systems were technically functioning. The gap existed in the space between them, in the handoff between a physical action and a digital record.

That gap didn't appear because someone failed to maintain a control. It appeared because the process design assumed a level of integration that didn't exist. The warehouse workflow and the ERP workflow were two separate things that were supposed to happen in a particular order. They didn't, reliably. And no control in either system could see that the sequence had broken down.

That's a systemic condition. Fixing the specific control finding doesn't close the gap. It addresses the symptom while leaving the condition that produced it intact.

The honest assessment of most material weakness remediation programs is that they're reactive, category-specific, and local. They fix the finding in the place where it was found. They almost never ask the broader question: what does this finding tell us about how our systems, processes, and people actually work together, and where else is that same dynamic creating exposure?

THE 18-MONTH LAG

There's a timing dimension to all of this that deserves its own attention.

Walk through the lifecycle of a typical audit finding. A control fails at some point during the fiscal year. Let's say it happens in the second quarter. The business keeps running. The failure either isn't visible yet, or people are working around it without realizing it.

The fiscal year ends. The audit begins, typically several months into the following year. The auditors come in, scope their work, test their samples. They find the issue. Fieldwork concludes. The report gets drafted, reviewed, revised. The audit committee is briefed. The final report is issued.

By this point, you might be eight to twelve months past when the control actually failed.

Then remediation planning starts. Management responds to the finding. Remediation plans get developed, reviewed, approved. Resources are allocated. Work begins. If the finding is a material weakness, the remediation timeline is under scrutiny. Even under scrutiny, closing a material weakness typically takes at minimum one full audit cycle to demonstrate.

So from the time a control fails to the time it's remediated and tested, you're easily looking at 18 months. In some cases, longer.

During that entire period, the business has been making decisions, approving transactions, recognizing revenue, recording expenses, and reporting results. All of it happened under a control environment that had a known gap that nobody knew was there yet.

The auditors themselves describe this limitation without always naming it directly. Audit scope is defined months in advance, based on prior-year findings and materiality assessments. Tests run on samples, often 25 to 50 items drawn from populations of millions. The work is conducted on a fixed calendar. By the time the report is issued, the finding reflects something that happened in the past.

This doesn't make audits useless. External auditors validate the integrity of financial reporting. Internal auditors provide independent assurance over operations and controls. That work has genuine value and the audit profession performs it rigorously.

But the audit model was designed for a world that moved at human speed, where transactions happened in batches and systems were updated periodically. In that world, an annual audit with a 90-day fieldwork window and a 60-day reporting cycle was reasonably well-calibrated to the pace of the business.

That world is gone. Enterprise systems now run continuously. Transactions process in milliseconds. Access changes propagate across systems in real time. A control failure that would have taken months to cause visible damage in 1995 can produce a material outcome in days now. The speed of the business has changed. The audit cycle hasn't kept pace.

The 18-month lag isn't a failure of auditors. It's a structural mismatch between the audit model and the operational reality of the modern enterprise. The lag was always there. It just matters more now.

WHAT GETS LOST IN TRANSLATION

There's one more thing audit reports don't tell you, and it's subtle enough that it rarely gets named directly.

The report describes what the control environment looked like during the audit period. It doesn't describe what the control environment looks like today. By the time you're reading it, the organization has changed. People have left. New systems have been deployed. Configurations have been updated. Business processes have shifted. A finding that was valid during fieldwork may have already been partially addressed through changes made for unrelated operational reasons. Or it may have gotten worse because of a system change nobody connected to the original control concern.

The report is a document about the past. Most conversations around it are about the present.

That mismatch creates its own kind of exposure. Management discusses findings in the context of their current understanding of the business, not in the context of the period the audit was testing. Audit committees ask questions about remediation, drawing on the implicit assumption that the environment being fixed is the one that exists today. It usually isn't quite the same environment the auditors tested.

Skilled practitioners on both sides understand this and mentally adjust. They track what's changed. But the process itself doesn't formalize the adjustment. The report is the report. The current state is the current state. Formal process often treats them as one thing.

This is part of why the shift from periodic to continuous assurance matters so much. The question isn't only how fast you find problems. It's how accurately the picture you're working from reflects the business you're actually running right now. An audit report tells you what the controls looked like at a point in time. A continuous assurance model tells you what the controls look like today, whether they're drifting, and where they're likely to break next.

The distinction sounds technical. The consequences are financial.

WHAT TO DO WITH A FINDING

If the six categories are symptoms and audit reports describe the past, what does good remediation actually look like?

It starts with asking a different question after a finding lands. The typical question is: what do we need to do to close this finding? A more productive question is: what does this finding tell us about where our systems, processes, and people aren't aligned, and where else is that misalignment showing up?

That second question is harder. It requires looking across systems rather than within one. It requires understanding the business process end to end, not just the control point where the failure was detected.

When a reconciliation finding surfaces, the control answer is to tighten the reconciliation. The systemic answer is to trace back to where the data actually diverges. Is the source system producing the right output? Is the data transformation in the integration layer accurate? Are there manual interventions happening upstream that aren't being captured anywhere? The reconciliation failure is often the last visible symptom of a longer chain of events. Fix only the last link and the chain will break again somewhere else.

When an access control finding surfaces, the control answer is to revoke the excessive access and redesign the role. The systemic answer is to ask how that access accumulated in the first place. Was it provisioned incorrectly at the start? Did it expand through role modifications over time? Did a merger or acquisition bring over an access model from a different system without proper review? Those answers determine whether the fix holds or whether the same access creep reappears in six months in a slightly different form.

When ITGCs are cited, the specific remediation matters less than the broader question: what controls depend on this ITGC infrastructure, and should any of that evidence now be treated with skepticism? A weak ITGC environment doesn't just produce a finding. It potentially taints the reliability of everything that depends on that infrastructure for its evidence base.

Three things to do immediately after receiving an audit report:

Map each finding to the business process it touches, not just the control category. Write out the process from start to finish and mark where the finding occurred. Then look at what sits upstream and downstream from that point.

Ask whether the same underlying condition could be producing findings in other categories that weren't tested in this audit cycle. SoD findings in one system often correlate with access control issues in adjacent systems. Reconciliation failures often connect to integration gaps. Change management weaknesses often affect ITGC evidence quality. Look for the cluster, not just the individual finding.

Take the finding out of the audit cycle and assign a business owner, not just a control owner. The difference is real. A control owner fixes the control. A business owner asks whether the process design needs to change. Most recurring findings need the second conversation, and most organizations skip it.

The audit report is the start of a conversation, not the full picture. The categories it names are real, but they're the places where failure becomes visible, not the places where failure begins. Treating them as root causes rather than signals produces remediation that's expensive, time-consuming, and temporary.

The most useful thing you can do with an audit report is read between the lines. What the auditors found tells you something. What they didn't look for tells you more.

Chapter 4
How Good Tools Create Bad Outcomes

Somewhere in your organization right now, there's a GRC tool running a segregation of duties analysis. There's a security platform scanning for suspicious access patterns. There's an audit management system tracking evidence requests. There's an identity governance tool processing role changes. There's a SIEM correlating log events. And there are almost certainly a few spreadsheets bridging the gaps between all of them.

Every one of those tools is doing its job. Every one of them was bought for a real reason, probably after something went wrong or an auditor asked a question nobody could answer cleanly. And yet the risk picture you're looking at is still incomplete.

That's the strange reality of enterprise risk right now. The more competently you've solved individual problems, the harder it gets to see the whole. You haven't failed by being careless. You've arrived here by being careful, one purchase at a time.

A BRIEF HISTORY OF POINT SOLUTIONS

To understand how we got here, it helps to trace the sequence of decisions that made sense at the time.

Start with the ERP itself. When large enterprises implemented SAP in the 1990s and early 2000s, the promise was integration. One system of record. Finance, procurement, manufacturing, HR, all running on a common platform. And to a meaningful degree, that promise delivered. The core transactional controls inside SAP S/4HANA and its predecessors are genuinely strong. Authorization objects, workflow approvals, change documents, posting controls, table logging. The digital core was built to be controllable.

The problem wasn't the ERP. The problem was everything that grew around it.

As enterprises got bigger and regulations got more demanding, the gaps inside the ERP became visible. Segregation of duties was getting violated, but nobody could see it systematically. So vendors built tools for that. SAP GRC Access Control came along and gave organizations a way to analyze SoD conflicts, manage role requests, control firefighter access, and govern emergency authorizations. It was a real solution to a real problem. This spurred additional competition from multiple vendors over the years to attempt to manage strategy over SAP's authorization concept.

Then auditors wanted evidence that controls were actually operating, not just designed correctly. So process control tools emerged. SOX compliance programs created demand for control documentation, testing workflows, and sign-off tracking. SAP GRC Process Control filled part of that gap for SAP environments. Platforms like MetricStream, RSA Archer, and AuditBoard addressed it for the broader enterprise.

At the same time, identity and access sprawled beyond what any ERP could govern on its own. Cloud apps arrived. Contractors and third parties needed system access. The lifecycle of a user's permissions got complicated fast. Identity governance platforms like SailPoint stepped in. Azure AD and Okta took on the authentication layer. SAP released its own Identity Access Governance solution for cloud environments, and the legacy SAP Identity Management system was still running in many shops.

Then cyber threats escalated and log data became critical. Security information and event management tools, the SIEM category, grew into an industry of their own. Splunk, QRadar, and others promised to make sense of the enormous volume of system events that enterprise environments generate. SAP Enterprise Threat Detection showed up specifically for SAP-native log analysis.

Third-party specialists filled narrower gaps. Xiting, Pathlock, SecurityBridge, and Onapsis each built deep expertise in SAP security and controls and IT general controls, offering capabilities that the native tools didn't cover well. Each of them solved something specific very well.

And behind all of this, the cloud application layer kept expanding. SuccessFactors for HR. Ariba for procurement. Concur for expense management. Fieldglass for contingent labor. Each with its own control model, its own access governance approach, its own audit footprint.

So by the time you zoom out and count what a large enterprise is actually running today, the list is striking. According to patterns across the client environments I've worked in, most large enterprises run somewhere between four and six disconnected control tools at the ERP layer alone, and that count grows substantially when you add identity, security, and cloud application monitoring to the picture. One study of the market put the average number of point solutions per enterprise at four to six times higher than it was just five years ago.

Every single one of those tools earned its place. Every purchase decision made sense in context. And now you're sitting on a stack of individually excellent systems that produce collectively incoherent output.

THE TOWER OF BABEL PROBLEM

Here's the part that doesn't get talked about enough in vendor briefings or audit committee meetings.

Each of those tools was built by a different team, with a different data model, speaking a different control language. When they produce output, they produce it in their own format, using their own terminology, mapped to their own version of what "controls" means.

Your SoD tool talks about role conflicts and mitigation approvals. Your process control platform talks about control objectives and test results. Your identity governance system talks about entitlements and certifications. Your SIEM talks about events, anomalies, and alert severity. Your audit management system talks about findings, recommendations, and remediation status.

None of these languages map to each other without human translation.

That translation work falls on your people. Every quarter, someone on the audit team or in the controls function has to pull exports from four or five different systems, build a spreadsheet that manually reconciles the outputs, and try to construct a coherent picture of where things stand. That work consumes somewhere in the range of 40 to 50 percent of the average audit team's time, based on the patterns I've seen consistently across different environments. Nearly half of your most senior controls resources, spending their time on data assembly rather than risk analysis.

The Tower of Babel analogy is accurate and deliberate. In the biblical story, the builders weren't incompetent. They were skilled. The problem was that they stopped being able to understand each other. When the language broke down, the work broke down with it. Progress was impossible not because the workers lacked capability, but because the shared context was gone.

That's exactly the situation in most enterprise risk functions today. The individual contributors are skilled. The tools are capable. The data exists. But there's no common language, no shared context, and no one systematically connecting the outputs. Each team speaks its own dialect, and the risk that matters most hides in the spaces between those dialects.

Consider a scenario that plays out in some form across many

large enterprises. A user in the finance department gets a new role approved in the identity governance tool because the role itself appears clean. That same user already has an existing role assignment in SAP that, when combined with the new one, creates a meaningful SoD conflict. The GRC access control tool would flag that conflict if it ran a full analysis, but it runs on a different schedule and hasn't processed this combination yet. Meanwhile, the process control platform is showing green on the relevant financial controls because the controls themselves are documented and signed off. Three tools, all operating correctly, none of them seeing the same picture. The conflict exists. Nobody's seeing it.

This is the Tower of Babel problem in practice. It's not that the tools failed. It's that the tools weren't built to talk to each other, and nobody assigned the job of making them do so.

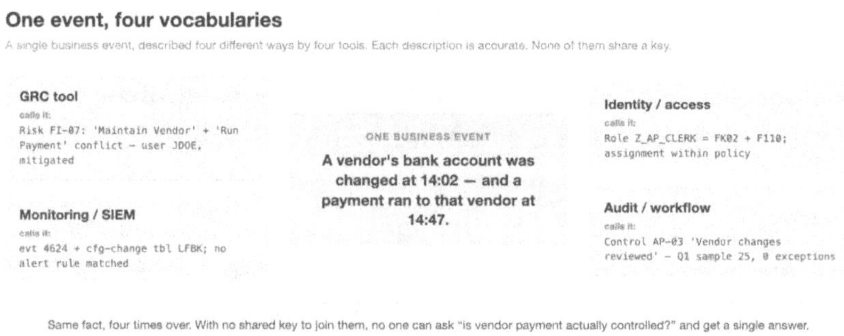

THE NORMALIZATION OF FRAGMENTATION

What makes this genuinely dangerous is how thoroughly the industry has accepted it as normal.

If you walk into almost any large enterprise today and ask the CAE or the CRO how they're managing control fragmentation, you'll get one of a few answers. Most commonly, you'll hear about a manual reconciliation process. Someone coordinates. Someone compiles. Someone reviews. The answer is usually a person, not a system.

Often it's a very capable person who has been doing it long enough to hold the institutional memory in their head.

When that person leaves, the knowledge leaves with them. And the organization discovers, usually during the next audit cycle, just how dependent they were on one individual's ability to translate between systems that don't speak the same language.

This fragmentation got normalized for understandable reasons. Budget cycles favor solving the most visible problem in front of you. An auditor flags a SoD issue, you buy a SoD tool. A breach reveals an identity gap; you buy an identity governance platform. A SOX deficiency surfaces around evidence management; you buy an audit management system. Each decision is rational. Each decision is made in isolation. The cumulative effect is a stack of tools that collectively create a new class of risk they weren't designed to address.

Organizational structure reinforces it. The GRC team owns the access control tool. IT security owns the SIEM. Internal audit owns the audit management platform. The identity team owns the IGA solution. Each team has a different reporting line, a different budget owner, and a different definition of what "risk" means in their context. Cross-tool coordination requires cross-functional cooperation, and cross-functional cooperation in a large enterprise requires deliberate design. Without that design, the default is each team stays in their lane.

Vendor incentive structures make it worse. Every vendor in this space is motivated to make their tool more capable, not to make their tool more interoperable with competitors. Deeper capability within a silo is a feature. Compatibility with adjacent tools is, at best, a checkbox. So the tools get better at what they do, the interfaces between them stay weak, and the gap widens.

The result is that most enterprises have normalized a situation where their risk posture requires human interpretation to understand, can only be assembled after the fact, and depends on people rather than systems to hold together. When you spell that out plainly, it sounds alarming. But because it's been the standard for so long, it reads as acceptable operating procedure.

Compliant and exposed, as I said earlier, is the condition. Normalized fragmentation is the mechanism that produces it.

THE COMPETENCE TRAP

There's one more layer to this that I want to name directly, because it's the one that tends to surprise senior leaders when they see it.

The problem gets harder to solve as the individual teams get better at their tools.

Think about what happens in a mature SAP GRC program. Over years of work, the team has built deep expertise. They know the rule sets. They know how to tune the SoD analysis. They've built custom reports. They understand the quirks of the authorization model and how different role combinations behave. They're genuinely excellent at what they do.

That expertise is real and valuable. It's also a trap.

Because when you've built that much institutional knowledge around a single tool, the tool becomes the frame through which risk gets interpreted. The GRC team sees risk as SoD conflicts and mitigation approvals. They're not wrong. They're looking at one genuine dimension of risk. But they're not looking at timing. They're not looking at behavioral anomalies. They're not looking at whether the controls are actually executing when they're supposed to. They're looking at access design.

Meanwhile, the security team has built equivalent expertise in their environment. The identity team has built equivalent expertise in theirs. Each team has optimized its view. Each team is correct within its frame. And because each team has invested years in becoming very good at their specific tool, the tool becomes the end in itself. Keeping the tool configured, running, and producing output is the mission. Whether the output connects to a complete risk picture is someone else's problem.

The competence trap works like this: the more skilled each team becomes at managing their silo, the more the silo deepens. The skill creates confidence, the confidence reduces perceived urgency around integration, and the integration gap grows invisibly while everyone is busy being excellent.

I've watched this play out across a lot of organizations. A control function that runs a textbook SoD program, with clean workflows and well-documented mitigations, sitting right next to an identity governance operation whose certifications and role recertifications look pristine. Both programs would hold up well to scrutiny in isolation. But when you look at the combination, user access lifecycles and SoD profiles are moving in different cycles, on different schedules, with different exception processes. The hand-off between the two programs is manual and infrequent. A user who should be flagged in both systems is flagged in neither, because neither team has assigned anyone the job of checking the intersection.

It's not incompetence. It's two competent teams doing exactly what they were hired to do, with nobody responsible for the space between them.

This is why the answer to fragmentation is rarely more capability inside any single tool. The tools are already capable. What's missing is the discipline of orchestration, the function of connecting those competent, siloed outputs into something coherent.

That discipline doesn't exist inside any vendor's product roadmap. It's not the audit team's job, or the GRC team's job, or the security team's job. Which is why, until someone names it and assigns it, nobody owns it.

The field you're sitting in right now has an enormous competency in the pieces and almost no competency in the whole. Changing that requires understanding first why the pieces are so good at staying separate, which is what this chapter is about.

Understanding is step one. Step two is seeing what the whole actually looks like when you map it out. That's where we go next.

WHERE THIS LEAVES YOU

Before moving on, it's worth taking stock of what the fragmentation actually costs in concrete terms.

The direct cost is time. If roughly half of your audit team's hours go toward manually reconciling control evidence across systems, that's half your most expensive risk talent doing data assembly. It's not analysis. It's not insight. It's transcription.

The indirect cost is confidence. When your risk picture requires manual assembly, there's always a lag between what the systems know and what you know. That lag was measured in weeks during the annual audit cycle model. As automation accelerates and transactions happen at machine speed, a weeks-long lag in your risk visibility becomes a structural exposure, not just an operational inconvenience.

The strategic cost is credibility. When your board or audit committee asks whether controls are working, the honest answer in most fragmented environments is: "We believe so, based on what each system reported last time we ran a full reconciliation." That's a fundamentally different answer from "Yes, and here's the current evidence." One of those answers belongs to a CFO or CAE running a modern risk function. The other belongs to a function that hasn't changed its operating model since 2004.

None of this happened because organizations made bad decisions. It happened because the industry solved problems the same way it always does: one product at a time, one budget cycle at a time, one team at a time.

The next step is understanding what it actually looks like when you pull all of those pieces into a single view. Most organizations have never mapped their full control landscape. When they do, what they find is usually surprising and almost always clarifying.

That mapping exercise, and what it reveals, comes a little further on. But first, the next chapter looks at what happens when organizations try to buy their way out of fragmentation — and why a pile of tools, however good each one is, still isn't a strategy.

Chapter 5
A Pile of Tools Is Not a Strategy

You cannot buy your way to coverage. By the time most enterprises notice the fragmentation we have been describing, the instinct is to fix it with a purchase, and the vendor they turn to first is usually the one that already runs their core. To its credit, SAP has built an answer for almost every risk you can name.

Look at SAP's own governance and risk portfolio today and you find roughly ten separately licensed products. Access Control analyzes segregation of duties and governs provisioning. Process Control and the newer Financial Compliance Management handle continuous control monitoring and testing. Risk Management runs the register. Business Integrity Screening hunts fraud and anomalies. Watch List Screening checks business partners against sanctioned-party lists. Cloud Identity Access Governance extends access control to cloud applications. Enterprise Threat Detection watches the platform for attacks. Secure Login handles authentication. UI Data Protection masks and logs sensitive data on screen. Each is a capable product. Each solves something real.

Now read the price tag — not in dollars, but in what it takes to own all of it. Each product carries its own license. Several require a

Business Technology Platform tenant to run. At least one, Watch List Screening, is metered by transaction volume and depends on a separate content subscription from a third party. Each needs its own implementation project, its own configuration, and its own specialists to keep it tuned. SAP markets the set as a cohesive, modular suite across the three lines of defense. It is cohesive. It is also, in full, a multi-year, multi-million-dollar program.

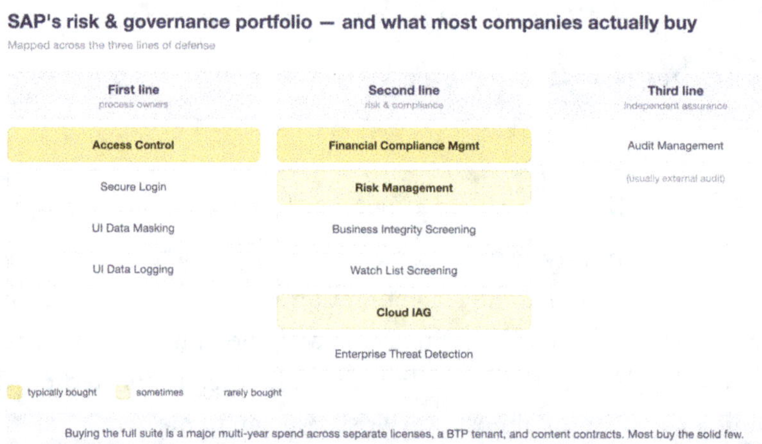

What organizations buy instead is a subset, and the subset is rarely chosen by strategy. It's chosen by pressure. Access Control gets bought because SOX forces a segregation-of-duties answer. Process Control sometimes follows when the auditors want evidence of monitoring. Business Integrity Screening shows up after a fraud scare. The rest — threat detection, data masking, watch-list screening, the cloud governance layer — stays on the shelf until an incident or a regulator makes it urgent.

This is the same reactive purchasing pattern we saw earlier, now playing out inside a single vendor's catalog. Coverage accumulates one finding at a time. The result is an estate over-covered in the places auditors look hardest — access and segregation of duties — and blind in the places they don't: the integration layer, cross-application conflicts, and the third line of defense that most companies hand to external audit entirely.

Consider one seam that no checklist catches. A security tool flags a suspicious technical event on an account; separately, a controls tool lists that same account as holding a segregation-of-duties conflict. Apart, they are two low-priority tickets in two different systems. Together, they are an emergency — and nothing joins them.

You don't have to take my word that this pattern is real; it is sitting in the public filings. Strip the names and the same ecosystem failure shows up again and again — and in every case the signal that would have caught it already existed in one system, while the awareness only ever comes from reading it against another.

What the Disclosures Were Really Saying

Composite cases drawn from public SOX 404 material-weakness disclosures (FY2019–2024).

What happened (anonymized)	Signal that already existed	Where it should have been captured	The cross-reference that raises real awareness
Specialty retailer: ERP access not restricted, duties not segregated, system changes undocumented.	User-to-role assignments creating SoD conflicts; production changes with no approval record.	Access-risk analysis (SoD) and the change / transport log.	The SoD conflict, read against the event log that shows it was exercised and the undocumented change record — one user who both holds the conflict and used it.
Biotechnology: within one ERP module, the same access could change data, and the changes weren't monitored.	Master-data change documents posted by privileged users, never reviewed.	Change-document and table logging.	Who can change, read against who did change and the approval evidence — exposing maintain-and-monitor in one pair of hands.
Communications, post-acquisition: inherited ERP with SoD, program-change, journal-entry, reconciliation and period-end weaknesses.	Journal entries posted by conflicted users; uncontrolled program changes; open reconciliations.	GRC SoD and change management on the inherited ERP.	The journal entry, read against the poster's access, the approval workflow and the reconciliation status — across the inherited and corporate estates at once.
Global manufacturer: two country sites running local systems outside the corporate ERP, each with SoD conflicts.	Access conflicts on the local systems, invisible to the central GRC tool.	Local-system access review (never centralized).	The central SoD ruleset extended to the local systems and read against their transaction logs — the local exception becomes visible to the center.
Consumer retail: ITGC weakness in an operational inventory system that feeds inventory valuation.	Access and change to the warehouse system, and its interface into the ERP.	ITGC over the operational system — usually scoped out as "not finance."	The inventory-system change, read against the interface posting into the ERP and the valuation impact — tying an operational system to the financial statement.

Examples are illustrative composites drawn from public U.S. SEC Form 10-K, Item 9A disclosures of material weaknesses in IT general controls — segregation of duties, user access, and change management — for fiscal years 2019 through 2024. Company identities are withheld; the underlying disclosures are a matter of public record.

None of these was a single broken control. Each was a control-ecosystem failure that ran in the same sequence: an access weakness opens a segregation gap, the gap allows an unreviewed change, the change makes a process control unreliable, and the evidence to prove otherwise was never assembled — so the auditor, finding nothing, records a deficiency that later hardens into a material weakness. In every row the signal already existed; what was missing was the reading across systems in the final column. None of it required buying another tool.

The strategic error isn't that the company failed to buy enough SAP. It's that there was never a coverage decision at all. Spend followed findings. No one asked the prior question: given the risks that would actually hurt this business, which signals do we need to see, and which sources produce them?

That question reorders everything. Coverage stops being a procurement list and becomes a design. Start from the processes that carry real exposure — vendor payments, revenue recognition, the financial close. For each, name the risk that matters and the signal that would reveal it forming. Then check what you already own that emits that signal. Only then do you decide: buy, build, or connect what's already there.

Framed that way, most enterprises discover they don't need ten products. They need the few that produce the signals their real risks demand, and a way to read those signals together. That reading is the orchestration discipline this book is about. It is what lets you get a coherent answer from a deliberately chosen subset instead of an exhaustive and unaffordable whole. The more sources you have, the richer the picture — but deliberate beats complete, and a chosen handful read together beats a full shelf read separately.

A fair question at this point: doesn't SAP's own next-generation

GRC close this gap? SAP is unifying its governance products onto a single, in-memory platform and adding AI assistants that monitor controls and draft rules. That's a real advance, and it lowers the cost of integrating the SAP-native pieces with each other. But it doesn't change the coverage problem, because the coverage problem was never only inside SAP. Most enterprises run more than one ERP, a dozen non-SAP applications that touch money, and an identity and security stack that no single vendor governs. Unifying the SAP tools is necessary. It is not the same as seeing the whole estate.

None of this is an argument against SAP's products. They are good, and an effective orchestration layer reads from them — the more of them are in place, the more it can see. The argument is narrower and more useful: owning every tool a vendor sells is neither realistic nor a strategy. A strategy is the decision about which risks you will watch, which signals reveal them, and how you will read those signals as one picture. Tools are the inputs to that decision. They are not a substitute for making it.

THE MAP OF THE SEAMS

A scannable map of the six kinds of seam where risk hides between systems — vendor-neutral, so it stays accurate as tools change. The next chapter walks each one in depth.

The recurring sequence behind almost every finding: access weakness → SoD gap → unauthorized or unreviewed change → unreliable process control → missing evidence → audit finding → material weakness. Each seam below is one link in that chain that no single tool watches.

The Map of the Seams

The six kinds of seam where risk hides between systems — vendor-neutral, so it stays accurate as tools change.

#	The seam	The two fragments (one tool each)	What neither sees	The cross-reference that closes it
1	**Process-to-system**	System: authorization objects, posting rules. Process owner: the actual workflow.	Steps added after go-live, or shadow processes run outside the system.	Process map × system authorization × the path the transaction actually took.
2	**Cross-application SoD**	In-system SoD tool: conflicts within one app. Identity tool: entitlements per system.	A conflict assembled across apps — approve an onboarding in one system, change vendor bank details in another.	Access in App A × access in App B × whether one person holds both.
3	**Evidence-timing**	Monitoring and audit: periodic, sampled, backward-looking. The business: transactions settling continuously.	The unobserved window between a risk event and the next control review, while money becomes irreversible.	Transaction event × control-review window × the moment of irreversibility.
4	**Configuration-vs-intent**	Change management: what changed, one approved change at a time. Risk and controls: what controls should exist.	Accumulated drift away from the original design intent.	Original design intent × current configuration state.
5	**Tool-to-tool translation**	Each tool: an accurate answer in its own language. The CFO: is this process controlled right now?	No common language to assemble one answer — it lives in a hand-built slide.	Normalize every tool's output to one shared control model.
6	**Security-meets-finance**	Security monitoring: a suspicious technical event. Controls: a financial SoD conflict.	A threat event on the exact account that also holds a financial conflict — two low-priority tickets apart.	Security event × the account's financial-conflict status.

How to read it: the risk never lives in a single row of a single tool. It lives in the last column — the join. Orchestration is the discipline of computing that join continuously, so the seam closes before the finding, not after.

Chapter 6
The Enterprise Landscape Nobody Sees Whole

Pull up the last risk report you received. Now ask yourself one question: does it show you the whole picture, or does it show you the picture one tool chose to paint?

Most senior leaders already know the answer. They've sat in steering committees where the GRC dashboard says green, the audit team is still chasing documentation, the identity system is flagging anomalies nobody has investigated, and the ERP is quietly processing transactions no one sanctioned. Each system is doing its job. Each system is telling part of the truth. Nobody in the room is holding all of it at once.

That's not a technology failure. It's a visibility failure. And the first step to fixing it is understanding what your actual landscape looks like — every layer, every tool, every gap between them.

A MAP OF WHAT YOU ARE RUNNING

Most organizations have never drawn a complete map of their enterprise control environment. They have vendor diagrams, architecture slide decks from two implementations ago, and system inventories that live inside IT. What they rarely have is a single picture that shows the financial systems, the cloud apps, the integration platforms, the GRC tools, the security stack, and the identity layer all in one place — with the control coverage mapped across all of it.

That missing picture is what this chapter is about.

Let's start with what a large enterprise actually runs. If you've spent time inside an SAP environment, this will feel familiar. If you haven't, it still translates, because the pattern is the same in virtually every major ERP estate.

At the center sits the digital core. In SAP terms, that's S/4HANA or the older ECC platform. It's the system of record for financial and operational truth. Finance (FI/CO), Order-to-Cash (SD), Procure-to-Pay (MM), Manufacturing (PP), Asset Management (PM), Quality (QM), Project Systems (PS), Treasury and Risk — these processes live here. Every major financial transaction originates, settles, or gets recorded in this system. The controls embedded here are transactional controls: authorization objects, workflow approvals, posting rules, change documents, table logging. They're well understood. They've been audited many times.

Around that core sits a ring of line-of-business cloud applications. These are the SAP products that have moved off the on-premise ERP over the past decade. SuccessFactors runs HR and payroll. Ariba manages sourcing and procurement. Concur handles travel and expense. Fieldglass manages the contingent workforce. SAP's CX tools cover sales and service. Each of these is a purpose-built system with its own data model, its own user access framework, its own control environment, and its own way of reporting what it knows.

They feed the core. And they each have their own version of the truth.

Outside that ring sits the broader enterprise. Salesforce for CRM. Workday for HCM if the organization hasn't committed fully to SuccessFactors. ServiceNow for IT service management and, increasingly, GRC workflows. Coupa for procurement if Ariba hasn't been adopted. Okta or SailPoint for identity and access governance. Bank connections and payment rails. Industry-specific platforms.

Data lakes built on Snowflake, Databricks, or BigQuery. Security tooling for threat detection and logging.

Connecting all of these is the integration layer. In modern SAP environments, that's SAP BTP and the Integration Suite. API management, event mesh, extension runtime, central logging. This is the plumbing that moves data between the digital core and the outer rings. It's where the enterprise is supposed to reduce fragility by decoupling systems from each other.

On top of all of this sits the governance, risk, and controls stack. SAP GRC Access Control for segregation of duties and user provisioning. SAP GRC Process Control for continuous controls monitoring. SAP GRC Risk Management for the risk register and response workflows. SAP IAG and Cloud Identity Services for identity governance. SAP Enterprise Threat Detection for security monitoring inside the ERP. Third-party tools like Xiting, Pathlock, SecurityBridge, and Onapsis for deeper SAP security analysis. AuditBoard, MetricStream, and RSA Archer for audit management and enterprise GRC. SailPoint and Okta for enterprise-wide identity.

Every one of those tools is legitimate. Every one solves a real problem. Most of them are well implemented in large organizations.

And none of them shows you the whole picture.

The enterprise landscape, seen whole

Every SAP and adjacent control system, drawn as one picture — the digital core at the center, the control applications reading into it, and one assurance layer reading across them all.

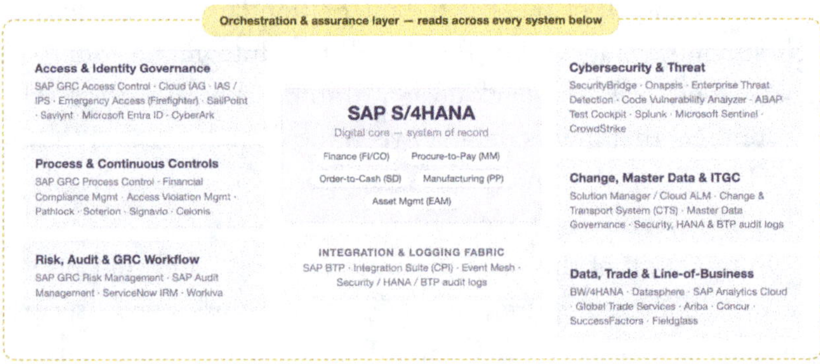

Every control system already feeds the core. What no system owns is the line around the outside — the layer that reads across all of them at once.

THE INTEGRATION LAYER AS BLIND SPOT

Here's what tends to surprise people when I walk them through this landscape for the first time: the integration layer is usually the least controlled part of the whole stack.

Think about what the BTP Integration Suite actually does. It moves data between systems. It transforms payloads, routes messages, triggers events, and executes API calls across the enterprise. A payment instruction that originates in a supplier invoice in Ariba flows through the integration layer into S/4HANA before it hits the payment run. An HR termination in SuccessFactors flows through the integration layer before it triggers deprovisioning in the identity system. A bank detail change flows through before the change takes effect in the vendor master.

These are high-value, high-risk transactions. And the integration layer is processing them at machine speed, with minimal human review, often with logging that nobody is actively monitoring.

The controls that sit on either end — in Ariba, in S/4HANA, in SailPoint — don't see what happens in the middle. They see what arrives at their door, but they don't see what was modified, delayed, duplicated, or rerouted in transit.

The integration layer is where the SAP source documents say one thing and the downstream system shows another. It's where timing discrepancies live. It's where a well-designed control on one side never catches an anomaly on the other side, because neither tool is looking at the middle.

I've seen organizations invest heavily in GRC and identity tooling on both ends of a critical process, only to discover that the integration platform sitting between them had open API endpoints, inconsistent logging, and no control owner. The controls existed. The gap was in what connected them.

This isn't unique to BTP or SAP. Every enterprise integration platform carries the same risk. The integration layer is the central nervous system of the modern enterprise. It's also the part of the landscape that most risk inventories leave blank.

FIVE GAPS THAT SHOW UP EVERYWHERE

After nearly three decades working inside enterprise risk and controls, I've found that the specific failures vary by company. But the categories of gaps are remarkably consistent. No matter the industry, no matter the size, no matter how many tools are deployed, the same five structural gaps appear.

The process-to-system gap. Controls are designed to protect business processes. But the systems that execute those processes were configured by people who understood the system, not always the process. When a new process emerges, such as a vendor onboarding step that got added post-go-live, or an emergency journal entry workflow someone built as a workaround, the control map doesn't get updated. The process changes. The system stays the same. The gap grows.

This is why you can have a technically correct authorization structure in SAP and still have process-level exposure. The authorization object says the user can't approve their own purchase order. But if the purchase order workflow bypasses the ERP because someone built a shadow process in SharePoint to get approvals done faster, the authorization control never fires.

The process is the risk. The system is just a window onto part of it.

The cross-application SoD gap. Segregation of duties conflicts that exist within a single system are detectable. SAP GRC Access Control was built specifically to find them. But SoD conflicts that span applications are almost never systematically checked.

Consider a scenario that isn't unusual. A user in SuccessFactors has the role to approve a new employee onboarding record. That same user has a role in S/4HANA that allows bank account modifications for vendor master records. Those aren't the same system. The SoD tool checks each system separately. Nobody checks the combination.

Now add the identity governance layer. SailPoint or Okta can enforce access policies within their scope. But they're not mapping the business-process risk of combining a SuccessFactors approval with an S/4HANA vendor master change. That cross-application visibility requires someone to connect those two data sets manually, or build a query across systems that most organizations have never built.

Most enterprises have a clean SoD report for SAP. They have a separate, equally clean report for their identity platform. Neither report shows what you actually need to know.

The evidence-timing gap. Controls testing is almost universally backward-looking. An auditor selects a sample from the last quarter. A monitoring tool runs a nightly or weekly report. A compliance team reviews exceptions from the previous month.

The underlying transactions happen continuously. A payment run executes at 2:00 AM. A configuration change goes through at 11:58 PM on a Friday. A journal entry gets posted on December 31st with an effective date of December 1st. By the time any of these show up in the control evidence layer, the transaction is already settled, the configuration is already in production, and the journal entry is already in the financial statements.

The timing gap isn't just a convenience issue. It's a structural exposure. In the chapter on what audit reports don't say, we walked through how the material weakness disclosures often describe symptoms. The evidence-timing gap is one of the underlying causes. Controls generate evidence. That evidence gets reviewed. But the window between when a risk event occurs and when the control review catches it is measured in weeks or months. During that window, the business is effectively unobserved.

The configuration-versus-intent gap. Every ERP system gets configured at go-live to reflect how the business intends to operate. Then the business changes. Reorganizations happen. New product lines get added. Subsidiaries get acquired. The ERP configuration gets adjusted by project teams, BASIS administrators, and sometimes power users with change authority. Each individual change is small. Approved, documented, moved through the change management process.

But configuration drift accumulates. Six months after go-live, the authorization design still reflects the original organizational structure. A year later, the posting rules may have been relaxed to accommodate a workaround that was supposed to be temporary. Three years later, the system is running with a configuration that diverges from the original design intent in dozens of small ways, none of which triggered a control finding on its own.

No single tool tracks this systematically. Change management tools track what was changed. Risk management tools track what controls exist. Security tools track who can do what. Nobody is holding the original design intent up against the current configuration state and identifying where they no longer match.

The tool-to-tool translation gap. This is the gap that makes the other four harder to fix. Every tool in the landscape produces its own output in its own format. SAP GRC reports SoD violations in terms of authorization objects and risk IDs. AuditBoard reports control test results in terms of control names and test conclusions. SailPoint reports access exceptions in terms of entitlements and policy rules. SecurityBridge reports threat events in terms of attack patterns and severity scores.

Each of those outputs is technically accurate. Each one is useful to the team that runs that tool. But when a CFO asks whether a specific business process, say vendor payments, is actually controlled right now, the answer requires someone to manually translate from SAP GRC's language, plus SailPoint's language, plus the process documentation, plus the last test result in AuditBoard, plus any open findings in the GRC risk register.

That translation is usually done in a PowerPoint by someone who's pulling from memory and last month's reports. It's the best available approximation of an answer. It's almost never a complete picture.

The tools don't speak to each other. And the enterprise risk profile lives in the conversation between them.

WHAT EACH TOOL SEES – AND WHAT IT MISSES

It helps to be specific here. When I say each tool sees only part of the picture, I don't mean they're poorly built. These are mature, often excellent products. What I mean is that each one was designed to solve a specific problem, and that scope boundary is baked into how it works.

SAP GRC Access Control sees role assignments, authorization object values, and the SoD ruleset. It knows who has what access inside SAP. It doesn't know whether that access is being used in combination with access in Ariba, Concur, or a third-party platform. It doesn't know whether the person who has payment-approval authority in SAP also happens to be the one who approved the vendor's creation in the vendor onboarding portal.

SAP GRC Process Control sees automated controls that run against SAP data. It can tell you whether reconciliations are being completed, whether posting periods are closed correctly, whether manual journal entries above threshold are getting reviewed. It doesn't see controls in non-SAP systems. It doesn't cross-reference its results with the identity layer or the audit workpaper to confirm the whole process is covered.

AuditBoard and similar audit management platforms see what auditors have tested and what conclusions they've reached. They hold the workpapers, the evidence, the open findings, and the remediation status. They don't see the live control state in the ERP. A control can be documented as effective in AuditBoard because it passed last quarter's test, while the underlying system configuration that made it effective has since changed.

SailPoint and Okta see identity and access across the enterprise. They know who has access to what systems, whether that access was properly provisioned, and whether it's been certified recently. They don't see what those people are actually doing with that access. Access certification says the access is appropriate. It doesn't say the combination of accesses across systems creates a business-process risk.

SecurityBridge and Onapsis see threat events and vulnerability conditions inside SAP at the platform and security level. They flag unusual login patterns, Basis-level changes, and known attack techniques. They don't contextualize those events against the business process. A Basis-level change to the payment configuration is a different risk depending on whether it happened during a scheduled maintenance window with proper authorization, or at 3:00 AM by a user with emergency firefighter access that was never deactivated.

ServiceNow GRC holds the risk register, the control framework mappings, and the policy library. It's where management attests to control effectiveness. It doesn't consume live system data. The attestation reflects what management believes to be true, based on evidence they've reviewed. It doesn't update itself when a configuration changes or a new SoD conflict gets introduced.

Each tool is doing its job. Each job is a piece of the whole. The enterprise risk picture requires all of them speaking to each other. Right now, in most organizations, they don't.

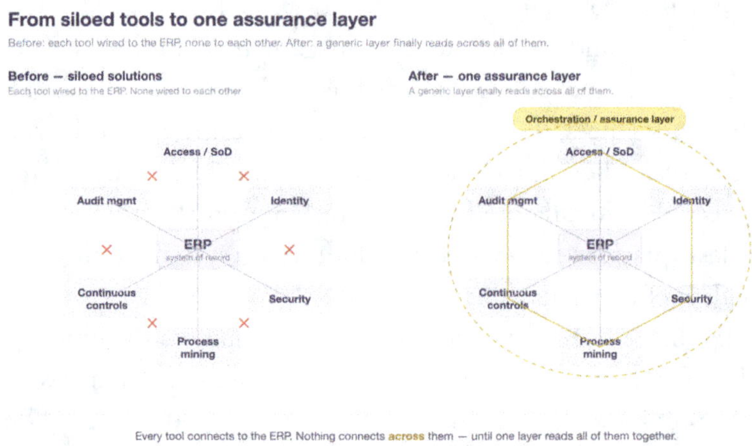

TWO VIEWS, ONE CONVERSATION NOBODY IS HAVING

There's a useful mental model for this landscape, borrowed from how large organizations use enterprise architecture diagrams. The same environment can be drawn two ways.

One is the functional landscape — what the organization owns and why, in terms a board or CFO can read. The other is the integration architecture — where systems depend on each other, where data transforms, and where the control gaps actually hide. Most organizations hold these two views separately and never lay them side by side.

The risk doesn't live in either view. It lives in the relationship between them — in the seam where what one diagram shows is exactly what the other conceals. We'll come back to how you build both views, and use each with the right audience, when we get to building the picture. For now, the point is simply that one diagram is never enough.

WHY THE MAP HAS TO COME FIRST

Some of this might feel like foundation-setting before the real work. It is. But it's foundation-setting that most organizations skip.

The standard approach to enterprise risk is to start with the framework — COSO, COBIT, SOX controls, whatever the regulatory requirement demands — and work down from there to systems and evidence. The framework defines what you're supposed to have. The audit tests whether you have it. The GRC tool tracks the results.

That approach produces a compliance picture. It tells you whether you satisfy the framework requirements. It doesn't tell you whether your actual technology landscape, with all its integrations, configurations, cloud apps, and third-party connections, is operating the way you intend.

To get from compliance to assurance, you have to start the other way. Start with the landscape. Understand what you're actually running. Understand where each process executes, which systems it touches, where control accountability sits, and where the evidence lives. Then map that against the framework. Then identify the gaps.

That's a different exercise. It's harder. It requires people from risk, IT, finance, and operations to sit in the same room looking at the same picture. It requires someone to own the question: do we actually know what this landscape looks like? In most digital transformations, the business process procedures are never fully documented — even though documenting them is one of the most critical steps in an ERP implementation. Without that documentation, the gaps may never surface, even with the best intentions.

In my experience, the honest answer is usually no. Not fully. The digital solution core is well documented. The GRC tools are well understood. The integration layer is a partial map. The cloud apps are individually documented but not shown in relationship to each other. The third-party connectors are somewhere in an IT inventory that hasn't been updated in eighteen months.

Nobody has drawn the whole thing. Nobody has layered the control environment on top of it. Nobody has identified, explicitly, what each tool sees and what it misses.

That exercise alone surfaces risk findings. Every time I've walked an organization through it, things come up that no individual system had flagged. A cloud HR system with no formal access review cycle. An integration endpoint with logging disabled. A third-party application with sensitive financial data that feeds the ERP but has never been included in the scope of the IT general controls audit.

The landscape exists whether you've mapped it or not. The risk exists in its gaps whether you've named them or not. The map doesn't create the risk. It makes the risk visible.

Visible is where you have to start.

WHAT TO DO BEFORE THE NEXT AUDIT

You don't need to wait for a project to begin building this picture. Three things are immediately actionable.

First, pull your current system inventory — from IT, from finance, from the GRC tool if one is running — and identify every application that touches a financial transaction, a people record, or a control process. Include the integration platforms. Include the cloud apps. Include the third-party tools. You're looking for scope, not detail. The goal is to see what you're working with.

Second, for each major business process where you carry material risk (vendor payments, payroll, revenue recognition, inventory, financial close), trace the process across the systems it touches. Write it out: which system initiates the transaction, which systems it passes through, which system records it, and where the control checkpoints are. Do that for five processes and you'll have identified more cross-system gaps than your last three audits combined.

Third, ask your audit team one question: which parts of this landscape are currently outside your testing scope? The answer will tell you where your assurance coverage has structural holes. Not because auditors are doing their jobs wrong, but because audit scope is bounded, and every scope boundary is a place where the controls could be drifting without anyone noticing.

Those three steps don't require new tools. They require someone with the authority to convene the conversation and the persistence to hold the whole picture together long enough to see what's in it.

The landscape nobody sees whole is not hiding. It's sitting in separate reports on separate dashboards in separate systems. The work is integration, not investigation.

And it starts with finally agreeing to look at the whole thing at once.

Chapter 7
Naming the Discipline

The chapters so far have built a case. The problem has a shape now, even if that shape hasn't appeared on a board slide yet. Organizations have good tools, solid teams, and reasonable intent. They still end up exposed. The gap isn't in their controls. It's in the coherence between them.

Before we build anything, though, we need language. Specifically, we need a name.

WHY NAMING MATTERS

There's a specific kind of problem that exists long before anyone names it. People feel it. They work around it. They spend money on it. They lose sleep over it. But because it doesn't have a name, they can't address it directly. They can only address the symptoms.

That's exactly where enterprise risk has been for the last two decades.

I've sat in dozens of post-incident reviews over the years. The pattern is always the same. Something went wrong. The audit found it afterward. Everyone in the room knew, in hindsight, that the signals were there. A GRC tool flagged an SoD conflict. The monitoring platform showed unusual posting activity. The access review hadn't been completed on time. But no one connected those signals into a single picture, because no one owned that picture. No one had a job title, a framework, or a budget line for connecting them.

When you don't name a problem, you can't assign accountability for it. You can't build a practice around solving it. You can't train people in it, fund it properly, or measure whether you're getting better.

Naming matters because it converts a vague, shared frustration into a solvable problem.

Think about what happened when "data governance" got named as a discipline. Companies had always had data quality problems. Inconsistent master data, duplicate records, competing definitions of the same metric. But it was background noise. Once data governance became a named field with its own frameworks, roles, and benchmarks, organizations could actually invest in it and track progress. The CDO role emerged. Standards developed. The field matured.

The same arc played out with information security. Before it had a name and a C-suite owner, organizations had IT staff patching servers and hoping for the best. Once it became a named discipline, the CISO role emerged, standards proliferated, and boards started asking real questions about it.

Enterprise risk management followed the same path. Once ERM had a name, COSO published frameworks, audit committees started asking for risk registers, and companies hired chief risk officers.

Each of those fields grew from the same starting point. Someone named the gap, gave it language, and built a practice around solving it.

The gap we've been describing throughout this book has been present for years. It's expensive. It's getting more dangerous as automation accelerates. And it hasn't had a name.

Until now.

THE FORMAL DEFINITION

Enterprise Risk Orchestration is the discipline of cross-referencing and blending the outputs of existing control systems into a single, coherent, continuous picture of whether the enterprise is actually in control.

Read that again, because each word is deliberate.

"Discipline" means a repeatable practice with frameworks, roles, skills, and accountabilities. It's not a software feature. It's not a project. It's not something you do once during an audit cycle. It's an ongoing operating capability that the organization builds and maintains.

"Cross-referencing and blending" means the work is integrative. You're not replacing your GRC tool or your identity platform or your SAP security layer. You're drawing on all of them at the same time. You're asking what you can see when you look at the output of three different tools together, and whether the combined picture matches what management believes is happening.

"Existing control systems" matters for two reasons. First, it makes clear you're not starting over. Most organizations have already invested significantly in point solutions. Those investments don't go to waste. Second, it sets the boundary. The discipline is about connecting what you have, not buying something new to replace it.

"Continuous picture" is the shift from the old model. Traditional GRC gives you a periodic report. This discipline gives you a live view. The difference is roughly the same as the difference between a photograph and a camera feed.

"Whether the enterprise is actually in control" is the destination. The goal isn't documentation. The goal isn't a clean audit opinion. The goal is a genuine, evidence-backed answer to the question every CFO and CRO is really asking: are we actually running this business the way we think we are?

That's the formal definition. It governs three things specifically:

business intent (what the organization is authorized to do and is actually trying to achieve), time-to-intervention (whether controls can act before outcomes become irreversible), and authority (who and what is actually permitted to make which decision, in the systems as configured, not just in the policy documents).

Those three dimensions show up in every major control failure I've seen. The policy said one thing. The system was configured to allow something slightly different. A transaction moved faster than any human could catch it. And nobody was looking across all three at once.

That's the gap Enterprise Risk Orchestration closes.

WHAT IT IS NOT

Defining a new discipline precisely requires saying what it isn't, because the risk of confusion is real. Senior leaders who hear "orchestration" will map it to things they already know. Some will hear GRC 2.0. Some will hear "another platform." Some will hear a consulting engagement. None of those are right.

Enterprise Risk Orchestration is not another GRC tool.

Your organization probably has at least one GRC platform already. Maybe two. Those tools do valuable work: they maintain control libraries, track risk registers, manage policy workflows, and collect evidence for auditors. They're good at what they do. But they operate within their own boundaries. They don't reach into your identity management system to cross-check access against SoD policies. They don't compare your SAP posting activity against your process mining output to see if the transaction sequence matched the authorized workflow. They work in their lane.

Orchestration works across lanes. It's the connective layer above the point solutions, not a replacement for any of them.

It's also not a monitoring tool, though monitoring is part of what the discipline enables. Monitoring watches for events. Orchestration asks whether the combination of events across multiple systems tells a coherent story, and whether that story matches what management believes is happening.

It's not an audit methodology, though it makes audit dramatically easier. Audit produces a periodic opinion. Orchestration produces a continuous state. The output of orchestration feeds the audit; it doesn't replace the audit's independence or judgment.

And it's not a vendor category that already exists, even though several vendors will claim it does once this language gains traction. If you search today for what this discipline describes, you'll find tools that touch pieces of the problem. GRC platforms, identity governance tools, process mining platforms, SIEM systems, and control testing automation. What you won't find is a mature, named practice for connecting all of those outputs into one coherent view, with defined roles, frameworks, and governance. That's the gap this discipline fills.

There's one more thing it isn't, and it's worth saying directly: it isn't the problem of any single team to solve alone.

The reason fragmentation persists isn't incompetence. It's ownership. The identity team owns access. The GRC team owns the control library. The process mining tool belongs to the operations group. The SIEM sits in cybersecurity. No one wakes up in the morning accountable for the coherence of what all four of those tools say together. That ownership gap is as important as the technical integration gap, and we'll come back to it shortly.

THE CONDUCTOR METAPHOR

The best way I've found to explain what Enterprise Risk Orchestration actually looks like in practice is a metaphor, and it's one that holds up under scrutiny.

Think of a large orchestra. You have strings, brass, woodwinds, percussion. Each section is technically excellent. The violinists have trained for years. The horn players are skilled professionals. The percussion section knows exactly what it's doing. If you put any one of those sections in a room alone, they'd produce something recognizable. Maybe even impressive.

Put them all in the same concert hall without a conductor, and you get noise.

The conductor's job isn't to play an instrument. It's to make sure every section is playing the right part, at the right tempo, in the right key, at the right moment. The conductor holds the full score. Each musician sees their own part, but only the conductor sees how all the parts are supposed to fit together, and whether they're actually fitting together right now.

Enterprise risk looks exactly like this.

Your SAP GRC tool is one section. It knows its part. It tracks SoD rules, monitors access risk, and flags violations. It's good at what it does.

Your identity and access management platform is another section. It governs provisioning, certifications, and role assignments.

Your process mining tool watches transaction flows. Your SIEM watches system events. Your financial controls platform reconciles accounts. Each one is technically sound.

Nobody is conducting.

Nobody holds the enterprise-wide score that shows how all those parts are supposed to fit together. Nobody is watching, in real time, whether the identity system's access grants are consistent with what the GRC tool says should be restricted. Nobody is comparing the transaction sequence the process mining tool sees against the control objectives the GRC tool has documented. Nobody has a picture of the whole orchestra.

Enterprise Risk Orchestration is the conductor's function.

Five excellent sections. Nobody holding the score.

Each tool is a section of the orchestra — technically superb on its own part. The conductor's job is to hold the whole score. Right now that podium is empty.

A conductor doesn't play an instrument. The role is to make every section perform the same score, at the right moment — the one job no tool is assigned.

To extend the metaphor a little further: the sheet music is your framework, whether that's COSO, SOX, COBIT, or your own internal control matrix. The instruments are your individual controls. The musicians are your point solutions and the teams who run them. The conductor brings all of that together in real time and hears the whole sound, not just one section.

A conductor doesn't replace the musicians. A conductor doesn't rewrite the music. The conductor makes sure the whole ensemble performs what the music says, when it's supposed to, in the way it was intended. When the first violin drifts out of tempo, the conductor catches it. When the brass comes in early, the conductor corrects it. That real-time awareness and correction, across the entire ensemble, is exactly what's been missing from enterprise risk.

The metaphor also captures something important about authority. A conductor has standing to tell any section to adjust. That standing comes from owning the score and being accountable for the performance, not from being the best musician in the room. The orchestration function in an enterprise similarly needs standing to reach across the GRC team, the identity team, the SIEM team, and the financial controls team. That standing has to come from somewhere. It has to be chartered.

WHO OWNS THIS DISCIPLINE

A discipline without an owner doesn't get practiced. It gets talked about in steering committees, included in strategy documents, and then quietly absorbed into whoever had space on their calendar.

Enterprise Risk Orchestration needs an explicit owner. The question is who.

In most organizations right now, no single function is positioned to own this naturally. Internal audit has the cross-functional view but typically lacks the real-time operational reach and the authority to change configurations or escalate in-flight findings. The CISO owns the security monitoring layer but often doesn't have the financial controls depth or the GRC framework fluency to connect the full picture. The CFO's office owns the financial controls and the SOX program but often doesn't have visibility into the identity management or cybersecurity layers. The CRO, where one exists, is often working from a risk register that's three steps removed from system-level evidence.

Each of those functions holds a piece of the conductor's score. None of them holds the whole thing.

What I've seen work is a small, cross-functional orchestration function that reports to the CFO or CRO and has explicit read access and integration rights across the relevant systems. It doesn't have to be large. Three or four people with the right skills and the right authority can do what no automated tool can do alone: hold the whole picture and ask whether it's coherent.

The skills this function needs are specific. Deep ERP knowledge, because most of the evidence lives in SAP or NetSuite and the ability to read what a transaction log or access log actually means is not generic. GRC fluency, because the orchestration function has to translate technical evidence into framework language that auditors and boards can use. Data integration capability, because the work is fundamentally about connecting outputs across systems. And something that's genuinely rare: the ability to look at signals from five different tools and form a judgment about what they mean together.

That last skill is what makes this a discipline rather than a software feature. Software can connect the data. Software can surface correlations. But the judgment about whether the pattern the data reveals is actually a risk, whether it represents a control that's drifted from its design, whether it indicates fraud or just sloppiness, requires a practitioner who's seen enough to know the difference.

Close to thirty years of building that pattern recognition is behind much of what this book argues. Part of what these chapters are doing is transferring enough of it to give you, and the people on your team, a sharper lens.

The ownership question also has a governance dimension. The orchestration function needs a clear escalation path. When it finds a coherence gap, when the GRC tool says access is restricted and the identity system shows it isn't, it needs a defined path to the CFO, the CRO, or the audit committee. That path has to be established before the finding, not after.

Organizations that treat this function as an advisory role with no escalation authority will find that the findings pile up in reports that nobody acts on. That's not orchestration. That's just more documentation.

There's also a board-level dimension. Audit committees are increasingly being asked whether the company has effective controls over financial reporting, over cybersecurity, over operational processes. The honest answer, in most organizations, is something like: we have effective controls over individual processes and we're fairly confident they're working in aggregate, but we don't have a systematic way to verify the coherence. That's the gap this discipline closes.

When the orchestration function is working well, the CFO and the CAE can walk into an audit committee meeting with evidence, not just attestation. The shift from "we have controls and we believe they're working" to "here is the live evidence that they're working, and here is where they aren't" is the core value this discipline delivers. It changes the conversation at every level of the organization.

The next chapter gives you the framework for how that function actually works: the four planes of assurance that structure the orchestration work and give it discipline rather than just good intentions.

PUTTING IT INTO PRACTICE

Naming the discipline is the beginning. The next question is how to start building the capability before you have the full architecture in place.

Three moves get you started before your next audit cycle.

First, create an inventory of your orchestration gaps. Pull the outputs from your five or six primary control systems, whatever those are for your organization, and map them against each other. You're looking for places where two tools should be telling the same story but aren't, and for places where a question that matters for control assurance can only be answered by combining data from more than one system. Most organizations find four or five significant gaps on the first pass.

Second, identify who currently comes closest to owning the coherence picture. In most organizations this is an informal role someone is already playing without a title or a budget. It might be a senior internal audit manager who has the deepest cross-system knowledge. It might be a control owner in the finance team who's been manually reconciling information across tools for years. Whoever it is, that person is your first orchestration resource. Formalize the role, give it standing, and protect some calendar space.

Third, run one cross-system question before your next executive reporting cycle. Pick one process, procure-to-pay is usually a good start, and ask: what does each of the tools that touches this process say about whether it's in control? Put those answers side by side in a single document. You'll almost certainly find at least one place where the answers don't match. That mismatch is your proof of concept. It's the concrete evidence you take to the CFO or the CAE to justify building the capability properly.

Those three moves won't build the full discipline overnight. But they establish the practice, name the owner, and produce a finding that makes the case for going further. That's enough to start.

Chapter 8
The Four Planes of Assurance

Every discipline needs a structure. Not a checklist you laminate and hang on the wall, but a real working model that tells you what to look at, in what order, and why it matters. Without that, Enterprise Risk Orchestration stays a good idea that never becomes an operating practice.

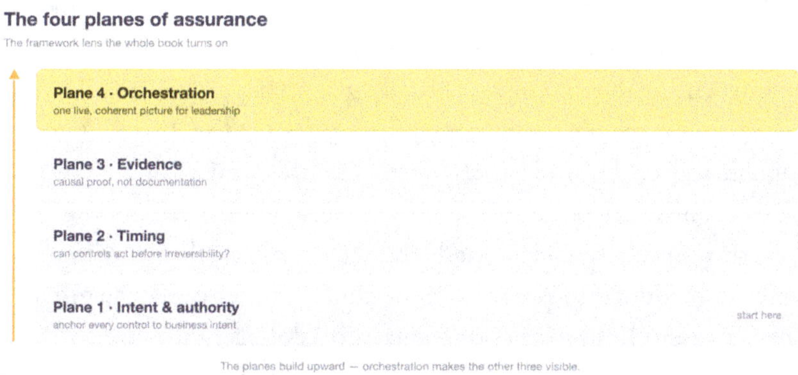

The four planes are that structure. They came out of nearly thirty years of watching risk failures up close, across clients whom engaged top tire consulting firms, and yet, at the time we didn't know they were failures because we were siloed focused. The pattern that kept showing up wasn't a lack of controls. It was controls that existed in isolation from each other, designed without reference to business intent, triggered on calendar schedules instead of risk moments, and supported by evidence that was documentable but not provable. The four planes are the answer to that pattern because consulting firms are still addressing risk in siloes. Those siloes are how clients are also structured and engage with their consultants. So both sides are actually creating and validating the siloed approach. Its time to restructure that strategy.

Each plane addresses a distinct dimension of the assurance problem. Together, they give you a complete picture of whether the enterprise is actually in control — not whether it passed a test, but whether it's operating inside the boundaries it intended to set.

A FRAMEWORK BUILT FROM THE FIELD

Frameworks usually arrive from two places: consultancies that build them top-down from first principles, or regulators that build them as documentation requirements. Both tend to produce structures that are logically sound and operationally awkward. They describe what good looks like without addressing how a real enterprise, with real systems and real constraints, actually gets there.

The four planes came from the opposite direction. They came from sitting in the middle of control failures and asking a consistent question: at what point did this break down? Was the intent wrong? Was the authority misaligned? Did the control fire too late? Was there no evidence to prove what actually happened? Or were all the pieces present but never connected into a coherent picture?

The answer was almost always one of those four things. Rarely one in isolation. Usually two or three at once, which is exactly why the failures were hard to catch.

So the framework is built to mirror how risk actually breaks. Each plane is a layer of the same assurance problem, and gaps in any layer show up as exposure in the business. You can have a perfect authority structure and still be badly exposed if your controls fire after irreversible commitments are made. You can have real-time evidence collection and still be unable to answer a basic audit question if the evidence isn't connected to the right framework requirement.

The four planes work as a system, which is why the last section of this chapter matters as much as the first four: reading them together is where the actual insight lives.

PLANE ONE – INTENT AND AUTHORITY

Start here, because everything else depends on it.

A control without a purpose isn't a control. It's a procedure. Procedures get followed until someone decides they don't matter, and in large enterprises, that happens more often than you'd think. The configuration gets inherited from the prior year's audit. The approval threshold stays where it was set during implementation. The segregation-of-duty rule stays mapped to roles that no longer reflect how work actually gets done. Nobody changed it because nobody owned it and nobody questioned whether it still served its original purpose.

That's what it means when a control is disconnected from its intent. The mechanism exists, but it's no longer anchored to the business outcome it was designed to protect.

The first plane fixes that. Every control in the assurance system has to be tied to a specific business intent: what the organization is trying to achieve, what's at stake economically if that control fails, and who or what is authorized to act within it. When the business changes, which is to say, constantly, the controls that were built around old decisions need to be recalculated against new ones.

That recalculation doesn't happen on its own. In most enterprises, there's no mechanism that connects a business decision, say a new acquisition, a new product line, or a new payment structure, to the control environment it touches. The acquisition happens. The integration plan focuses on systems and people. The controls that govern the combined entity get reviewed sometime later, usually during the next audit cycle, usually after something has already slipped.

The first plane creates that mechanism. It requires that controls be maintained not as static documentation, but as living definitions anchored to current business intent.

The second half of the first plane is authority. This is the part most enterprises think they've handled because they have role-based access control in their ERP and an SoD ruleset in their GRC tool. Those are necessary. They're not sufficient.

Authority, done properly, means defining exactly how much autonomy each role and each system is permitted, under what conditions, and with what escalation requirements. The distinction matters because modern enterprises don't just have human decision-makers. They have automated workflows, scheduled jobs, integration platforms, and increasingly, AI-assisted processes, all taking action inside the ERP without a human in the loop. Each of those systems has an effective authority level. The question is whether you've defined it or inherited it by default.

When you haven't defined it, you end up with situations that are technically within the access model but operationally outside the risk appetite. A system job that runs nightly with admin-level authority because that's how it was configured during implementation six years ago. A workflow that escalates exceptions to a queue that hasn't been actively monitored since the person who built the process left the company. These aren't access control failures in the traditional sense. They're authority failures: the organization never made a conscious decision about how much power to give these processes.

Authority profiles solve that. They're explicit, versioned definitions of what each role and each system is allowed to do, reviewed when the business changes, and connected to the evidence plane so you can see when authority was exercised and what happened as a result.

PLANE TWO – TIMING

Controls that fire after the damage is done aren't really controls. They're reports.

Most enterprise controls are designed around calendar schedules: monthly reconciliations, quarterly reviews, annual audits, SOX test cycles. That design made sense when business processes moved at human speed and the biggest risk was documentation errors caught at month-end. It doesn't match how enterprises actually operate today.

Consider what happens in a payment run. An employee or an automated process updates a vendor's bank account details in the ERP. The payment file transmits to the bank. The bank accepts and processes it. From the moment the bank account detail was changed, you had a window: a gap between the commitment and when a control could realistically intervene to stop or flag it. If your vendor master review runs monthly and the payment file runs weekly, the math isn't in your favor.

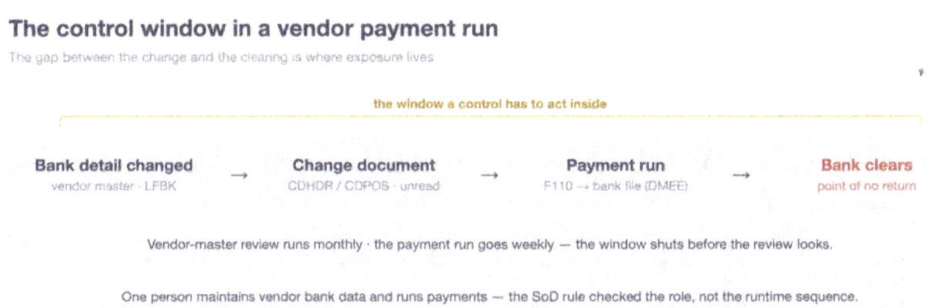

This isn't a hypothetical risk category. Vendor bank account fraud is one of the most consistent and costly control failures in enterprise finance. The pattern is always the same: the change was in the system, the control existed, and the control fired after the money moved.

The second plane is about closing that gap. It models the time window between when a business commitment is made and when controls can realistically intervene. The technical term is temporal risk, but the practical meaning is simpler: does your control respond before outcomes become irreversible?

In Source-to-Pay, the irreversibility moments are vendor bank detail changes, payment file transmission, and bank acceptance. Once the bank accepts a payment, you're in recovery mode. Your control window is everything that happens before that point.

In Order-to-Cash, the irreversibility moments are pricing overrides that affect margin, revenue recognition triggers that affect reported results, and credit approvals that commit exposure. Once revenue is recognized and reported, adjusting it is an accounting event with its own audit implications.

In Record-to-Report, the irreversibility moments are journal entry postings that affect financial statements, consolidation steps, and external filing triggers. Some of these are reviewable. But once you've filed, the bar for restating is very high, and the reputational cost is substantial.

Mapping these moments changes how you design and evaluate controls. A monthly reconciliation that covers a process with a two-day irreversibility window doesn't provide meaningful protection. It provides documentation. The two things are not the same.

When controls are evaluated against irreversibility moments rather than calendar schedules, investment in assurance starts going where it actually changes outcomes. You may decide to maintain a monthly control for a low-frequency, low-value process. You'll decide something very different for a daily payment run that moves material amounts. The plane gives you the analytical basis for that decision instead of just inheriting the schedule from last year's test plan.

A useful way to think about it: every significant business process has what you might call a reaction window, the time between when a risk event starts and when it becomes difficult or impossible to reverse. The control has to operate inside that window, or it's operating outside the risk zone entirely. Most enterprises have never mapped their reaction windows. They've mapped their test schedules.

There's a quieter version of the timing problem that never involves a payment file, and it produces more audit comments than any fraud scheme. It's the gap between when something happens in the physical world and when the system records it.

The warehouse ships on Friday afternoon to make the truck, but the goods issue isn't posted until Monday. The field technician closes the job on site, but the service entry sheet gets keyed at the end of the week. A production run is confirmed in a batch at shift change rather than as each unit completes. In every case the business event has already occurred; the ledger simply doesn't know yet.

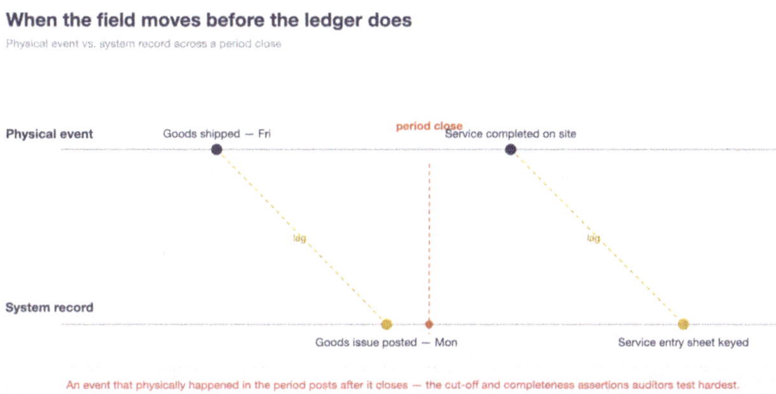

For a day or a weekend, that lag is invisible. At period end, it isn't. An event that physically happened in one period posts in the next, and now two of the assertions auditors test hardest — cut-off, did the transaction land in the right period, and completeness, did everything that happened get recorded — are both in question. This is how a field-to-ledger delay that nobody owned becomes a cut-off finding or a deficiency in period-end close controls. The control that would catch it isn't a tighter reconciliation; it's the orchestration that compares the physical-event stream against the posting stream and flags the divergence while the period is still open.

PLANE THREE – EVIDENCE

Evidence is the word that does the most work in assurance, and it's also the word that gets stretched furthest from its actual meaning.

In traditional GRC, evidence is mostly documentation: screenshots of system settings, PDFs of approval workflows, sampled test results, and policy sign-off sheets. These artifacts are useful. They show what was designed and sometimes what was tested. They don't reliably show what happened.

The distinction matters every time something goes wrong. When a control failure occurs and management asks what the system did, a screenshot of a configuration setting doesn't answer the question. It shows what the system was set up to do at a point in time. The question is whether it did it, when, under what conditions, and whether the outcome was what the business intended.

Traditional evidence can't answer those questions at scale. It was designed for a different era, when transaction volumes were lower, human reviewers were in the loop for more individual decisions, and the primary audience for assurance evidence was an annual external audit. In that context, sampling made sense and documentation was a reasonable proxy for actual behavior.

That context is gone. In a modern SAP environment, the processes that carry the most financial risk — payment runs, journal entries, vendor master changes, revenue recognition triggers — operate at machine speed, often overnight, often without a human in the loop for individual transactions. Sampling a few of those to test the control gives you a statistical comfort level. It doesn't give you evidence that the control operated on every transaction that mattered.

The third plane is about building evidence that's causal, not just documentable.

A causal evidence structure connects four things: the raw signal from the system (what actually happened), the control that was supposed to govern it (what was supposed to prevent or detect the event), the framework requirement it maps to (which SOX control, COSO principle, or internal standard this relates to), and the business impact if the control failed or succeeded. Every piece of evidence in the system carries that full lineage.

When evidence is structured this way, the questions you can ask change completely. You can ask which transactions exceeded authorization thresholds without human intervention before settlement. You can ask which controls have a response window shorter than the irreversibility point. You can ask which exception patterns look like a bypass pathway. These aren't audit walkthrough questions. They're queries against a structured evidence base.

The difference in speed is dramatic. A traditional SOX readiness process involves weeks of evidence assembly: collecting artifacts from different systems, reconciling formats, filling gaps, producing packages. The evidence exists in fragments across SAP GRC, the identity management system, the audit management tool, and a collection of spreadsheets. Putting it together is a manual project that happens once a year and consumes significant audit team capacity.

When evidence is graph-native, meaning each piece is connected to every related piece by traceable relationships, the assembly problem disappears. An external auditor's request becomes a query. An evidence package for a specific control becomes a pull, not a project.

There's also a truth-over-time dimension that traditional evidence misses entirely. A screenshot tells you the setting was correct when someone took the screenshot. A causal evidence graph shows you whether the control operated correctly over every relevant transaction across the entire period. That's the difference between showing the auditor you had a seatbelt and showing them you wore it for every trip.

Five types of evidence together tell a complete story. Declarative evidence establishes what was designed and documented. Operational evidence shows how systems are configured. Behavioral evidence shows what people actually did. Transactional evidence shows what the systems actually did, event by event. Economic evidence shows what impact resulted. Most enterprises have declarative evidence in reasonable shape. Transactional and economic evidence are the gaps where the real exposures hide.

PLANE FOUR – ORCHESTRATION

The first three planes give you the raw materials: controls anchored to intent, evaluated against timing, supported by causal evidence. The fourth plane is where those materials are assembled into a picture that leadership can actually use.

Orchestration has a specific meaning here. It's not aggregation, which is what most GRC tools do when they pull data from multiple sources into a dashboard. Aggregation puts things next to each other. Orchestration connects them, normalizes them to a common language, and surfaces the intersections where the real risk lives.

Consider what a CRO sees today. The GRC platform has a heat map of documented risks. The SAP security tool has a list of SoD violations and access exceptions. The audit management system has open findings from the last internal audit. The identity system has a report on privileged access. The process mining tool has flagged anomalies in the payment run. Each of these is a signal from a different instrument, in a different format, with a different severity scale and a different underlying methodology. None of them know what the others found.

The CRO's job is to synthesize those signals into a view of whether the business is actually in control. At most enterprises, that synthesis happens in a spreadsheet, or in a PowerPoint, or in a meeting where each function reports its own findings and leadership tries to figure out whether the collective picture is reassuring or alarming. It's a reasonable effort. It's also structurally late, structurally incomplete, and structurally dependent on whoever does the reconciliation work being both thorough and available.

The orchestration plane does that synthesis continuously. It sits across the existing tools without replacing them. SAP GRC keeps doing what it does. The identity platform keeps managing access. The audit system keeps tracking findings. The orchestration plane reads their outputs, normalizes them to a consistent control language, and assembles them into one coherent picture of enterprise risk posture.

What it shows management is fundamentally different from what any individual tool shows. It can surface questions like: a privileged access exception in the identity system overlaps with an unresolved SoD conflict in the GRC tool and three flagged transactions in the payment run. That combination, in the same process, in the same period, is a different level of concern than any one of those findings individually. No single tool would surface it. The orchestration plane surfaces it because it holds the full picture.

The practical output takes different shapes for different audiences. A board or audit committee sees a composite risk posture: are we within our risk appetite across the key processes, and where are the zones that need attention? A CFO or controller sees the financial control picture: is the integrity of the numbers defensible, and what's the evidence? A CRO sees the risk landscape: where are the active exposures, where are the trends moving the wrong direction, and what needs escalation? An external auditor sees an evidence package: complete, structured, traceable to source, and continuously current.

One of the most important things the orchestration plane does is end what I'd call framework fatigue. Most enterprises maintain separate control catalogs for SOX, for their internal standards, for any industry-specific regulatory requirements, and for whatever framework their external auditors prefer to work from. Each catalog needs its own mapping, its own documentation, its own test evidence. The overlap between them is substantial, but because they're maintained in parallel, the work is duplicated.

In the orchestration plane, frameworks are lenses, not schemas. The evidence is collected once, connected to controls once, and frameworks are applied as query overlays over that unified base. A new regulatory requirement, say a new SEC rule or an emerging standard, becomes a new lens, not a new catalog project. The evidence is already there. The interpretation of it changes.

The orchestration plane also serves as the trigger layer for the enterprise. When a risk threshold is crossed, when a timing gap is detected, when authority is exercised outside its defined parameters, the orchestration plane surfaces it and can route it to the appropriate remediation workflow: a ServiceNow ticket, a Jira task, an escalation to a control owner. The key architectural principle is that the orchestration layer reads from source systems. It doesn't write back to the ERP. That boundary matters because it keeps the integrity of the operational systems clean and makes the assurance layer auditable in its own right.

There's a word the access-governance world uses that deserves more suspicion than it gets: mitigated. When a segregation-of-duties conflict can't be removed, it gets a mitigating control assigned — typically a monthly review by some manager — and the conflict turns green. The trouble is that the mitigation is usually a record, not an activity. It was assigned once and has not been performed since. The conflict shows resolved precisely because it is mitigated.

A mitigated-and-ignored conflict is more dangerous than an open one, because an open conflict still nags and a mitigated one looks handled. Call it the companion to compliant and exposed: mitigated and unwatched. The fix is not a better mitigation record. It is to stop trusting the paper promise and watch the events instead.

Watching the events is where orchestration becomes concrete, and it runs as a chain. The access layer shows a user holds both halves of the conflict — maintain vendor bank data and release payment. On its own, that is only potential risk. The event log and change documents show the user actually exercised both sides; potential has become actual. The process-mining trace shows the transaction took a path that skipped the dual-approval step it was meant to hit. The approval that should back it, sitting in a service-management tool, either doesn't exist or is a four-second rubber stamp by a related party, in a system the ERP never sees. And the documents — the vendor-master change, the payment proposal, the invoice — make the finding evidenced rather than asserted.

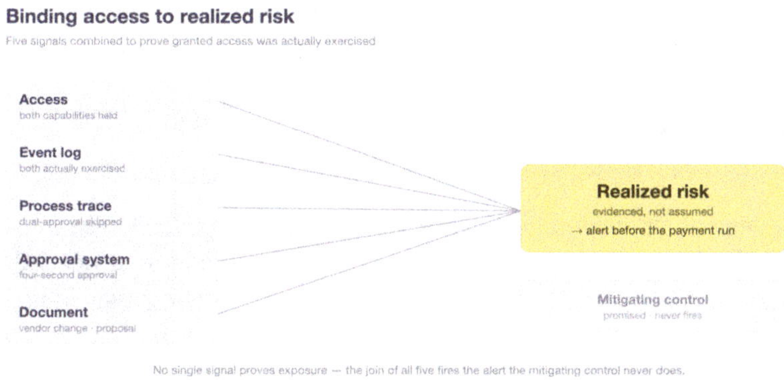

Link those five signals and you've made the move the entire framework is built toward: from this person could do something dangerous to this person did, the process confirms it, the approval was hollow, and here are the documents. That binding of access to realized, evidenced risk is the difference between an access report and a risk finding — and it is what a control looks like when it is genuinely monitored instead of merely mitigated.

READING THE FOUR PLANES TOGETHER

The power of this framework isn't in any single plane. It's in what you can see when all four are operating together.

Think about what a complete picture actually looks like. A control exists in the system. It's anchored to a specific business intent: protecting the integrity of the vendor payment process. Its authority profile defines that changes to vendor banking details require dual approval: one from the business owner and one from finance. Its timing model shows that any unapproved change made within 48 hours of a scheduled payment run represents a high-severity risk event, because the reaction window is too short for manual review to intervene before settlement. The evidence graph captures every vendor master change, every approval workflow execution, and every payment file transmission, with full lineage from raw event to framework requirement. And the orchestration plane holds all of that in context alongside the SoD picture from the identity system and the exception flags from the monitoring tool.

Now a vendor bank account changes on a Tuesday morning. Payment run is Thursday night. The approving manager's access was flagged as an SoD exception two weeks ago. The three signals, the change, the timing, and the access anomaly, appear together in the orchestration layer. Before Thursday's payment run, the relevant control owner has a ticket, a CRO has a dashboard alert, and the external evidence chain is already assembled.

That's a fundamentally different operating state than what most enterprises have today. Today, the bank account change might be in SAP. The SoD exception might be in the GRC tool. The payment schedule is in the treasury system. No one human holds all three at once unless someone builds a manual process specifically to look for exactly that pattern. That manual process, if it exists, runs on someone's calendar, not on the risk clock.

The four planes don't require you to replace your existing tools. That's important to say plainly. SAP GRC keeps managing access and SoD. The identity platform keeps managing provisioning. The audit management system keeps managing findings and test cycles. The monitoring tools keep generating their signals. The four planes sit over all of that and provide the coherence layer those tools were never designed to provide for each other.

There's also a maturity dimension worth naming. Not every enterprise can implement all four planes simultaneously. The right place to start is usually with Plane One: getting clear on which controls are actually anchored to current business intent and which have been inherited by default. That exercise alone surfaces significant gaps, and it doesn't require new technology. It requires a structured conversation between the risk function and the business.

From there, the planes build on each other. Timing analysis for the highest-risk processes. Evidence structure for the controls that govern those processes. And orchestration as the connective layer that makes the other three visible to leadership on a continuous basis.

The progression from there to continuous assurance is the subject of a later chapter. But the foundation is this: a framework that mirrors how risk actually breaks, built to run across the systems you already have.

The question most risk leaders ask at this point is a practical one: where do I start? The answer comes a few chapters on, where we build the landscape map that tells you exactly which processes carry the highest combination of impact and fragmentation, and therefore where the four planes will produce the most immediate and measurable return.

Chapter 9
Seeing the Whole Picture

Every conversation about enterprise risk eventually arrives at the same uncomfortable moment. Someone asks: "Can you show me, on one page, everything we're running and where the controls are?" The room goes quiet. People look at each other. Then someone offers to pull together a slide deck.

That slide deck never arrives. Or it arrives six weeks later, already out of date, covering about a third of the actual environment.

After nearly thirty years of walking into large enterprises, I've come to see that moment as one of the most honest things that happens in a risk conversation. The organization doesn't lack controls. It lacks a picture. And you can't manage what you can't see.

This chapter is about how to build that picture. We'll cover how to draw your landscape, where to look for your highest-risk gaps, why your control inventory is almost certainly incomplete, and how to use two different views to have two very different conversations with the right people.

DRAWING YOUR LANDSCAPE

The first step sounds obvious, but almost nobody does it well: write down everything your enterprise is running.

Not just SAP. Not just the systems IT officially supports. Everything. The platforms finance configured themselves. The middleware your integration team built three years ago that nobody documents. The SaaS tools your business units bought on a credit card before anyone knew procurement had a policy.

Start with four buckets. Your digital core is the center, usually SAP S/4HANA or an older ECC environment where the financial transactions of record actually live. This is your system of truth for revenue, payables, inventory movements, and journal entries. The controls here are well-understood: role design, authorization objects, workflow approvals, change documents, posting controls. It's the strongest part of your landscape, and most organizations know it reasonably well.

The second bucket is your line-of-business cloud applications. These are the SAP and non-SAP apps that connect to your core: Salesforce for the revenue pipeline, Workday or SuccessFactors for HR, Ariba or Coupa for procurement, Concur for expenses. Each one touches money. Each one creates or modifies data that eventually flows into your financial statements. Each one has its own access model, its own audit log format, and its own definition of what a "control" is.

The third bucket is your integration and platform layer. This is where SAP Business Technology Platform, middleware connectors, API gateways, and custom-built integration scripts live. This layer is almost always the most dangerous blind spot in the enterprise. It moves data between systems. It translates formats. It executes logic that nobody documented in a controls matrix. When something goes wrong in the gap between two systems, this is usually where it happened. As we saw when we mapped the enterprise landscape, the integration layer is where transactions lose their audit trail.

The fourth bucket is everything else: third-party infrastructure, security tooling, identity providers, data platforms, EDI networks, banks and payment processors. These aren't peripheral. They're the systems your ERP depends on to function correctly.

Write these four buckets down. Don't start with the ideal state. Start with what you're actually running today, as close to the truth as you can get it.

Once you have the inventory, draw the connections. Which systems send data to which other systems? Where does data flow from your operational apps into your financial core? Where does a transaction originate and where does it land? Don't draw every connection. Draw the financially significant ones, the ones where an error or an unauthorized change would eventually hit a financial statement.

This exercise usually takes a few weeks if you're doing it properly. It requires conversations with IT, with the integration team, with finance operations, and with the business. You'll discover systems people forgot existed. You'll find integrations that are "temporary fixes" from seven years ago. You'll find access paths that nobody thought to close because nobody thought to ask.

That's the point. Drawing the landscape is itself a risk activity. Every time you discover something you didn't know was there, you've found a potential gap.

FINDING YOUR HIGHEST-RISK GAPS

Once you have a landscape drawn, the next question is where to look first. You can't review everything simultaneously, and you shouldn't try. The goal is to identify the gaps where the distance between what the control is supposed to do and what the system actually did is largest.

Five types of gaps show up consistently across large enterprises. I've seen them across SAP environments of every size, in every industry.

The first is the handoff gap. This is the space between two systems where a transaction crosses from one to the other without a reconciliation checkpoint. A sales order closes in Salesforce. It triggers a delivery in SAP. The billing record is created, the revenue is recognized, and the cash lands in the bank. At any point in that chain, data can drift. The systems can disagree on timing, on amount, on customer identity, on terms. Most organizations have no continuous control sitting in that handoff. They have periodic reconciliations run by people, which find the discrepancies after the fact.

The second is the access gap. This isn't about having too many users with too much access, though that matters. It's about the distance between your access model in one system and your access model in another. A user might be terminated in your identity provider but still active in a downstream application. An SAP role might be technically SoD-compliant when analyzed inside GRC Access Control, but when you account for what the user can do in a connected cloud system, the conflict appears. No single tool sees both sides of that picture.

The third is the configuration drift gap. Systems change. Patches get applied. Configurations get adjusted to handle new business requirements. Someone turns off a control setting to troubleshoot a performance issue and forgets to turn it back on. Your IT general controls framework tracks some of this through change management processes, but it rarely tracks configuration changes in third-party SaaS tools. And it almost never tracks the combinations, what happens when a configuration change in one system interacts with a configuration in another.

The fourth is the process gap. This is the one that creates the most dramatic audit findings, because it's invisible to any tool. An order is cancelled in the order-management system, but the cancellation never reaches billing, so an invoice goes out and revenue posts for a sale that no longer exists. Finance closes the books using a number from one system while operations is still working from another. The purchase order is approved, but the goods receipt never matched. These gaps happen when the human process and the system process are out of sync, and no tool is watching both simultaneously.

The fifth is the evidence gap. This one is subtle and expensive. Your controls generate evidence: log files, workflow approvals, exception reports, reconciliation sign-offs. But who's verifying that the evidence is complete? That every required approval actually happened, not just that a box was checked? That the log file wasn't truncated? That the exception report didn't silently stop running when someone changed a batch job schedule? The evidence gap is where controls exist on paper but can't be proven in practice.

To prioritize which gaps to address first, score them across three dimensions. Start with financial materiality: if this gap allowed an error or manipulation, how large could the impact be? Next, assess the control coverage: how much detective and preventive control is already watching this area? Finally, consider velocity: how fast does a problem in this area move before someone notices?

High materiality, low coverage, and high velocity. That's your priority list. Work from there.

THE CONTROL INVENTORY YOU DO NOT HAVE

Here's something most organizations discover once they start drawing their landscape seriously: they don't actually know what controls they have.

They know what controls are documented. They know what controls they test for SOX. They know what their GRC tool is configured to monitor. But the actual control posture, across every system, process, and integration point, is almost always larger and messier than the documented version.

Some of that is good news. There are controls running that nobody wrote down, automated checks embedded in workflow configurations, system-enforced validations, default platform behaviors that prevent certain kinds of errors. These undocumented controls often carry real risk-reduction value. But because they're not in the inventory, nobody knows to test them, nobody knows when they stop working, and nobody credits them in the assurance picture.

Some of it is bad news. There are gaps where everyone assumes a control exists because it always has, but the system changed and nobody updated the configuration. The control is listed in the documentation. It's just no longer running.

The fastest way to understand your actual control inventory is to start from the business process, not from the tool. Take one of your major transaction cycles, order-to-cash, procure-to-pay, record-to-report, hire-to-retire, and map every step where financial risk could occur. At each step, ask three questions.

First: what is the control that's supposed to prevent or detect an error here? Second: what system is executing that control? Third: what evidence would prove the control ran and worked?

If you can answer all three questions with specifics, you have a documented control. If you can answer two out of three, you have a partial control that needs strengthening. If you can only answer one, or none, you have a gap.

Work through each step in the cycle. You'll find controls you forgot about. You'll find assumptions that don't hold anymore. You'll find places where two different teams each think the other one is responsible, which means effectively nobody is responsible.

This process is slower than running a configuration scan or pulling a GRC report. It's also more accurate. Tools can only tell you about what they're configured to see. Business process walkthroughs tell you what's actually happening.

Once you've done this for your highest-risk cycles, you'll have something more valuable than an audit workbook: you'll have a map of your actual control posture, grounded in evidence, showing you clearly where your coverage is strong and where it isn't.

That map is the foundation of everything else in this book. The four planes we covered give you the framework for thinking about controls. This inventory gives you the raw material to apply that framework to your specific environment.

A useful way to think about completeness is to test the inventory against the six categories where material weaknesses most commonly appear. Access controls. Segregation of duties. Journal entries. Reconciliations. Change management. IT general controls. For each category, ask whether you have a control identified, a system running it, and evidence proving it. If any of those three is missing for any category, that's a gap worth closing before someone else finds it for you.

Keep your control inventory in a form that can be updated. This is a living document, not a point-in-time snapshot. Controls change when systems change. Business processes change when the business changes. Your inventory needs to reflect both, which means someone needs to own it with an explicit responsibility to keep it current.

In most organizations, that ownership doesn't exist. The SOX team owns the documented controls. IT owns the system configurations. Operations owns the processes. Nobody owns the intersection. That's exactly where Enterprise Risk Orchestration lives.

TWO VIEWS, TWO CONVERSATIONS

Earlier, when we first mapped the landscape, I flagged that the same environment has to be seen two ways. This is where we build both. One of the most practical things I learned from working across Big-4 firms and large SAP environments is that the same enterprise landscape needs to be shown differently to different audiences. The picture doesn't change. The way you present it does.

There are two views that matter most. The first is a functional landscape view. The second is an integration architecture view. They're describing the same environment, but they're answering different questions, and they're designed for different conversations.

The functional landscape view answers the question: what do we own and why? It's organized by business function and domain.

Finance. Procurement. HR. Sales. Supply chain. Manufacturing. It shows which systems support which processes, and where the major dependencies are. This view is readable by a board member, a CFO, or a CRO who doesn't spend their days in system architecture diagrams.

When you sit down with a CFO and show a functional view of the enterprise, the conversation naturally goes to business risk. Where is revenue recognized? Which systems touch cash? Where are the vendor payment flows? What happens if the integration between order management and billing goes down? These are risk conversations in business language, and the functional view makes them possible.

The integration architecture view answers a different question: where is the risk and complexity? It's organized around data flows and system connections. It puts the integration layer at the center, showing how SAP connects to third-party SaaS, how identity systems feed into access models, how data moves from operational systems into the financial close. This view is designed for IT enterprise architects, security teams, and audit teams who need to understand where control failures can propagate.

When you show an integration architecture view to your internal audit team or your CISO, the conversation goes somewhere different. Where are the untested integration points? Which connections lack monitoring? If an unauthorized change were made in this middleware layer, which downstream controls would catch it? This is a technical risk conversation, and the integration view makes it productive.

Both conversations need to happen. Senior leaders need to see the functional view so they can connect the control picture to business outcomes. Technical teams need the integration view so they can identify and close specific gaps. If you're only having one of these conversations, you're missing half the picture.

The most sophisticated organizations use a combined view for transformation programs, when they're modernizing their ERP, adding cloud applications, or changing their integration architecture.

In transformation work, the question is: what changes when we modernize, and what controls need to move with it? That requires both lenses simultaneously.

For most risk and assurance leaders, the practical place to start is the functional view. Build it first, because it's the one that creates alignment at the top of the house. When a CFO or audit committee member can see the enterprise clearly enough to point to the areas of highest concern, you have an engaged sponsor for everything that comes next.

Then layer in the integration view for your technical teams. Give them the specificity they need to go find the gaps and close them. Let them work at the level of detail the functional view doesn't show.

The reason this two-view approach matters for orchestration is that different stakeholders own different parts of the risk picture. The CFO owns the financial exposure. The CISO owns the security posture. The CAE owns the assurance picture. The CIO owns the systems and integrations. Enterprise Risk Orchestration requires all four of them working from a shared understanding of the landscape. They'll never reach that understanding if they're each looking at a different slice of the picture and calling it the whole thing.

WHERE TO GO FROM HERE

Drawing the landscape isn't a one-time exercise. It's a discipline. The first time you do it is hard, because you're building from scratch and discovering things you didn't know. After that, maintenance is the job. Systems change. Business processes change. Integration points get added or modified. New SaaS tools get procured. The landscape you drew six months ago is already partially wrong.

The organizations that manage this well assign ownership at the intersection. Someone is responsible for knowing the current landscape, not just the documented one. That person works with IT, finance operations, the integration team, and audit to keep the picture current. They're not a system administrator. They're not an auditor.

They're an orchestration function, and in the AI era, that function becomes one of the most important roles in the enterprise.

For now, here's where to start.

Take one major transaction cycle, the one where financial risk is highest, and map it end-to-end. Identify every system that touches it. For each system, identify the controls in place and the evidence those controls produce. Find the handoff points where one system passes data to another without a continuous reconciliation check. Those handoffs are your first priority.

Then build the two views. Start with the functional view, since it travels to the conversations that create organizational alignment. Add the integration view once you have your technical team engaged. Keep both updated as your environment changes.

As you do this, you'll start to see the picture that your individual tools have never been able to show you. You'll see where your coverage is strong and where it's thin. You'll see where a single configuration change could create a gap that no existing tool would catch. You'll see where your evidence is solid and where it's assumed.

That's the picture most enterprises have never seen. And you can only manage what you can see.

Chapter 10
Risk as a Quality Discipline

In 1950, a statistician named W. Edwards Deming stood in front of a room of Japanese executives and told them something that sounded, at the time, slightly insane. Quality, he said, was not something you inspected into a product at the end of the line. You could not test your way to a good car by catching the bad ones before they shipped. Quality was a property of the process that built the car — and the only way to get it was to measure the process, find where it drifted from what it was supposed to do, change it, and then prove the change worked. Then do it again. Forever.

The executives listened. Within two decades, "Made in Japan" had gone from a punchline to a benchmark, and the companies that had taken Deming seriously were eating the lunch of the ones that hadn't.

I open with Deming because enterprise risk is sitting roughly where manufacturing sat before he showed up. We inspect. Once a year, sometimes once a quarter, we pull a sample of transactions, check them against the control design, and issue a verdict: effective, or not. We catch the bad ones we happen to catch. And then we release the pressure until the next inspection. What we almost never do is treat assurance as a process with a measured output that can be improved on purpose. We treat it as an exam to pass.

This chapter is about the move Deming made, applied to controls. It is, I think, the most important shift available to a risk function today, and it is the one almost nobody in the field is talking about. Everyone is selling integration — connect the tools, pull the data together, watch it continuously. That is genuinely useful, and the earlier chapters spent real time on why fragmentation hurts you. But integration by itself just gives you a cleaner dashboard. A cleaner dashboard is a better thermometer. It is not a quality system. The thermometer tells you the temperature. A quality system tells you what to change, and proves it worked.

WHAT DEMING ACTUALLY CHANGED

It is worth being precise about what the quality movement actually introduced, because the popular version — do things well, care about the customer — misses the mechanism.

Deming and the people who built on him — Juran, Ishikawa, the Toyota Production System that grew out of the same soil — introduced three specific things. First, a definition of a defect as a deviation from specification: a measurable gap between what the process was supposed to produce and what it actually produced. Second, a set of measurements that tracked that gap over time, so you could see drift before it became failure rather than after. Third, a loop — Plan, Do, Check, Act — that turned each measurement into a change and each change into a new measurement. The loop is the engine. Without it, measurement is just a more elaborate way of being told bad news.

None of that is about working harder or caring more. It is about building a system whose normal operation is to find its own gaps and close them, with evidence at every step. A factory running that way does not pass a quality exam at the end of the quarter. It is producing quality, and measuring that it is producing quality, continuously, as a byproduct of running.

Now hold that definition up against how most organizations assure their controls, and the mismatch is almost embarrassing.

INSPECTION IS NOT QUALITY

The annual audit is final inspection. It happens at the end of the line, after everything has already been built and shipped — after the payments went out, the access was granted, the journal entries posted. Samples get pulled. A verdict gets issued. And the verdict describes a period that, as we saw earlier, often ended six to eighteen months before anyone reads it.

Deming's whole point was that final inspection is the weakest, most expensive form of quality control there is. By the time you catch the defect, you have already paid to build it. You have shipped some number of defective units you didn't sample. And you have learned nothing about why the process produced the defect, so you will produce it again. He put it bluntly: cease dependence on inspection to achieve quality. Build quality into the process instead.

Here is the uncomfortable part. The continuous-monitoring tools the market is so excited about — the ones that pull every transaction and flag every exception in real time — are better inspection. They are a faster, more complete inspection line. That is a real improvement over annual sampling, and I don't want to wave it away. But more inspection is still inspection. A tool that flags ten thousand exceptions a day has not given you a quality system. It has given you ten thousand things to look at and no way to know which of them means your process is drifting, which fix would move the most risk, or whether last month's fix actually held.

A monitoring dashboard answers "what happened." A quality system answers a harder and more useful question: is the process that produces my controls getting better or worse, and what should I change first?

MEASURING THE GAP FROM SPEC

So what is the defect, in controls terms? What is the equivalent of "this part is three-tenths of a millimeter out of tolerance"?

It is the gap between what a control promises and what it actually does. Every control carries an implicit specification. The three-way match is supposed to stop a payment that doesn't reconcile to a receipt and a purchase order. The segregation rule is supposed to make it impossible for one person to both create a vendor and pay it. The change-approval workflow is supposed to ensure nothing reaches production without an independent sign-off. Those are the specs. The defect is the distance between the spec and the live behavior — the payment that went through anyway, the conflict that existed in practice, the change that shipped on a four-second approval that was really a rubber stamp.

Most organizations never measure that distance. They measure something easier and far less useful: how many controls they have. The control inventory is a count, and a count tells you almost nothing. Two companies can have the same three hundred controls on paper and be in completely different places — one where those controls are tight to spec, and one where half of them have quietly drifted into theater. The count is identical. The quality is not.

A quality discipline measures the distance, per control, and treats that distance as the number that matters. You can express it a few ways, and the labels matter less than the habit: How confident are you that this control is doing what it claims, and on what evidence? How far has its real behavior drifted from its design since the last time you looked? How fresh is the proof — is your confidence based on something that happened this week, or on a memory from last year's fieldwork? Those are quality measurements. They turn "we have a control" into "here is how well this control is actually performing, right now, against what it was built to do."

This is also, quietly, where the perceived-versus-actual gap from earlier in the book becomes a number instead of an anecdote. The whole problem this book has been circling — controls that pass on paper while the real exposure sits untouched — is, in quality terms, simply a population of controls running out of tolerance with no one measuring the deviation.

THE LOOP THAT CLOSES IT

Measurement on its own is inert. The Japanese factories did not win because they measured more; they won because every measurement fed a loop that produced a change, and every change fed back into the next measurement. Plan, Do, Check, Act. Applied to assurance, the same loop has four moves, and they should feel familiar, because they are how any honest investigator already works.

You observe what the controls actually did — not what the policy says they do, but the real behavior in the real systems. You compare that behavior to the spec and find where it diverges. You act to close the largest, most consequential gaps first. And then — this is the step that gets skipped — you confirm with evidence that the gap actually closed, and you carry that result forward so your picture of the control updates. Belief, action, evidence, revised belief. Then around again.

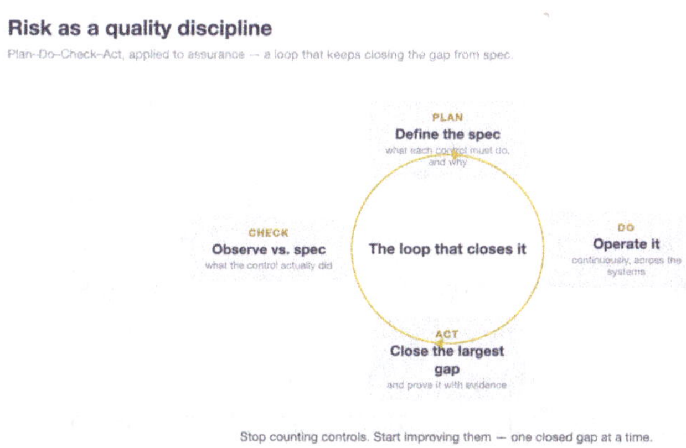

The step that gets skipped is the proof. It is genuinely common for an organization to identify a control weakness, launch a remediation project, hold status meetings about the remediation project, declare the remediation project complete — and never measure whether the underlying behavior actually changed. The finding gets marked closed because the project finished, not because the control started working. In a quality system, that is a category error. The project is the act. The proof that the defect moved is the check — and without it you have not closed the loop, you have just done some activity and assumed the best.

Run that loop continuously and something changes about the whole posture of the function. You stop lurching from exam to exam. The question is no longer "will we pass" but "is our control quality trending up or down this month, and is the thing we changed last month still holding?" That is a fundamentally calmer, more powerful position to operate from — and it is available now in a way it simply wasn't when evidence had to be gathered by hand.

FROM COUNTING CONTROLS TO IMPROVING THEM

Once you accept that assurance is a quality discipline, the maturity path the next chapter lays out reads differently. It stops being a ladder of "how often do we check" and becomes what it actually is: an improvement trajectory, in exactly the sense a quality engineer would recognize.

At the bottom, you inspect periodically and react to findings. A step up, you monitor continuously but still drown in exceptions, because monitoring without a quality frame is just faster inspection. Higher still, you are measuring the gap from spec per control, prioritizing by how much risk each gap carries, and closing the loop with evidence. At the top, the system is doing this well enough, and with enough history, that it can tell you where a control is about to drift out of tolerance before it does — the same way a well-instrumented production line flags that a tool is wearing before it starts cutting bad parts.

That top rung is worth pausing on, because it is the part that separates a quality system from a very good dashboard. A dashboard tells you a control failed. A mature quality system tells you a control is degrading — confidence slipping, evidence going stale, behavior creeping away from spec — while there is still time to act. Prediction here is not magic. It is what you get for free once you have been measuring the right gap, the same way a factory that has tracked tolerances for years can read the failure coming in the trend line. You cannot predict what you have never measured. The reason most risk functions can't see failures coming is not that the future is unknowable; it is that they have only ever counted controls, and a count has no trend.

WHAT TO PUT IN PLACE THIS QUARTER

You do not need to rebuild your control framework to start operating this way. You need to change what you measure, and add the loop. A few concrete moves, in rough order.

Pick your ten highest-stakes controls and write down their specs — in one plain sentence each, what is this control supposed to make true? You will find this harder than it sounds, and the difficulty is the point: a control no one can specify in a sentence is a control no one can measure.

For each, define the gap you would measure — what would tell you the live behavior has drifted from that sentence? Then go find that signal in the systems you already have. You are not buying anything yet. You are learning whether you can even see the deviation.

Stand up the loop on one of them, end to end: measure the gap, change something, then prove the gap moved. Doing it once, completely, on a single control teaches you more than a strategy deck does about ten.

And change one habit in how you close findings: stop marking a finding closed when the remediation project ends. Mark it closed when the measurement shows the defect actually moved. That single discipline — proof, not activity — is the smallest change with the largest effect, and it costs nothing.

None of this requires a new tool on day one. It requires a different idea of what the job is. The job is not to pass the inspection. The job is to run a process that produces controls you can trust, to measure how well it is doing that, and to make it better on purpose — with proof at every step. That is what Deming taught the factories. It is past time the risk function learned it too.

Chapter 11
From Periodic to Continuous

The audit calendar hasn't changed much in thirty years. Companies close their books quarterly, compile evidence packages in the weeks before fieldwork, sit with auditors for three to six weeks, then respond to findings that describe what happened six to eighteen months ago. The cycle repeats.

For most of that time, periodic assurance was the best available option. Evidence collection was manual. Testing was sample-based because nobody could test everything. Reporting took weeks because aggregating data from multiple systems required human hands at every step. The audit was periodic because the tools made anything else impractical.

The tools have changed. The process hasn't.

That's worth sitting with for a moment. The computing power inside a mid-sized enterprise today would have seemed like science fiction in the early 1990s. Systems run continuously. Transactions post in milliseconds. Logs capture every action, every exception, every configuration change. The data that could tell you whether your controls are working exists right now, somewhere in your environment. And most organizations still collect it once a year, in a sprint, under pressure, to hand to auditors who will review samples of it weeks later.

The result is a structural gap between when risk materializes and when anyone finds out.

THE ANNUAL AUDIT AS A REARVIEW MIRROR

Here's how audit preparation actually works at most large companies. About six to eight weeks before fieldwork begins, the controls team sends a request list to process owners. Those owners then spend two to three weeks hunting down screenshots, transaction logs, approval emails, and reconciliation sign-offs. Some of those documents are easy to find. Others require reconstructing what happened months ago from whatever trail someone thought to save at the time.

Then the auditors arrive, review samples from that evidence package, ask questions about things they didn't understand, and write findings. The draft report goes back and forth for another few weeks. Management responses are written, reviewed, and submitted. The final report closes the book on a period that ended roughly six months before anyone read the conclusion.

Six months is optimistic. For companies with complex environments or a large number of material weaknesses, the gap between when a control failed and when it shows up in a finding is often twelve to eighteen months. I covered the mechanics of that lag in the chapter on what audit reports don't say. The point here isn't to revisit how the lag happens. The point is that the entire model accepts the lag as normal.

Periodic assurance treats audit readiness as an event. You prepare, you get tested, you pass or you don't, and then the pressure releases until the next cycle. In between, controls run on whatever basis they were designed to run. Some work. Some drift. Some stop working entirely. You won't know which until the next event.

The rearview mirror analogy gets used a lot in risk circles, but I think it undersells the problem. A rearview mirror at least shows you what's immediately behind you. The annual audit shows you what was behind you a year ago. You're driving forward on today's road while getting real-time updates about what the road looked like last spring.

Nobody would accept that standard for anything else the business depends on. Cash positions get reconciled daily. Inventory levels get monitored continuously. Production lines trigger alerts when they deviate from spec. But the question of whether financial controls are actually operating as designed? That gets answered once a year, after the fact, in a document most executives skim.

WHAT CONTINUOUS ASSURANCE ACTUALLY MEANS

Before I explain continuous assurance, let me clear up what it's not. It's not running your audit every quarter instead of every year. It's not having your controls team test a sample every month instead of a sample every twelve months. Doing the same thing more frequently doesn't change the fundamental model. You're still sampling. You're still looking backward. You're still dependent on evidence that someone collected on a schedule.

Continuous assurance means the controls themselves produce evidence on an ongoing basis, and that evidence is monitored against known good states without waiting for a scheduled review.

The distinction matters. In the periodic model, evidence is collected for the purpose of auditing. Someone identifies what evidence is needed, goes and finds it, and gives it to a reviewer. The evidence exists only in the context of the audit. When the audit ends, the evidence package gets filed, and the system goes back to running without observation.

In the continuous model, evidence is a byproduct of how controls operate. Every time an access control fires, it generates a log. Every time a segregation-of-duty rule allows or blocks a transaction, that event is captured. Every reconciliation that runs produces a result. The evidence flows constantly because the systems are always running. The question is whether anyone is watching it.

Most IT departments already do some version of continuous monitoring. Security operations centers watch network traffic. Change management tools log every system modification. Authentication platforms track every login. The monitoring is there. What's missing is the layer that connects that monitoring to risk and control statements. Your SIEM knows that a privileged user accessed a production system at 2 a.m. Your ERP knows that a journal entry was posted outside normal business hours. But unless something connects those two events to the same person, to your segregation-of-duty policy, and to your financial close risk assessment, you've got two data points that nobody joins together.

Continuous assurance closes that gap. It takes the monitoring that already exists in individual systems and connects it to a common understanding of what controls are supposed to do, who owns them, and what the evidence of proper operation looks like. When a control drifts from its expected behavior, the deviation shows up immediately, in context, mapped to the risk it was designed to address.

The practical outcome is that audit readiness stops being a sprint and becomes a permanent operating condition. When external auditors arrive, the evidence package isn't assembled from scratch. It's drawn from a continuous record of control performance. Fieldwork gets faster because the evidence trail is already organized and timestamped. Findings shrink in number because issues that would have accumulated for twelve months are caught and corrected as they occur.

I've seen finance teams cut their audit preparation time significantly once this operating model is in place. Not because they're working harder in the weeks before fieldwork, but because there's nothing left to scramble for. The evidence was being collected all along.

THE MATURITY PATH

The shift from periodic to continuous doesn't happen in a single quarter. For most enterprises, it's a progression through distinct stages, each one building on the last. Understanding where you are on that path matters as much as understanding where you want to go.

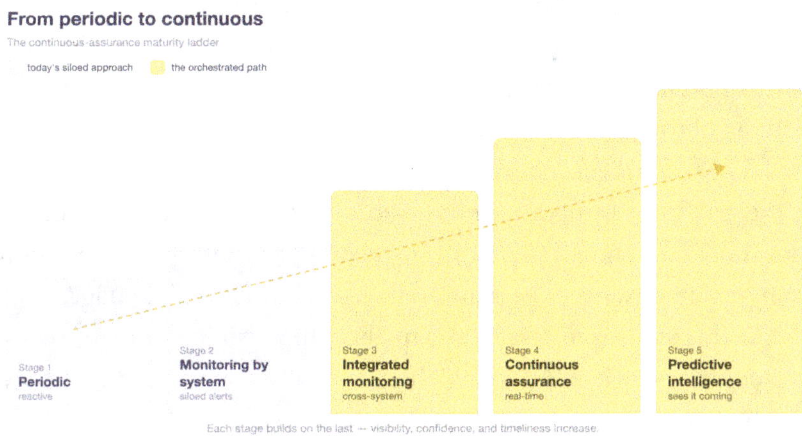

Stage one: Periodic. This is the current state for most organizations. Controls are documented, tested on a schedule, and reported in an annual or quarterly cycle. Evidence is collected reactively, usually in response to an audit request. The team is competent. The process is structured. But between formal testing periods, there's limited visibility into whether controls are actually running as designed. Issues surface when auditors find them, or when something breaks badly enough to become visible on its own.

If this describes your organization, you're not behind. This describes the majority of large enterprises right now. The problem isn't the people or the effort. The problem is the model.

Stage two: Continuous monitoring by system. Individual tools or platforms start producing alerts in real time. Your IAM system flags unusual access. Your ERP generates exception reports on segregation-of-duty violations. Your GRC platform tracks policy acknowledgments. Each system is watching its own domain. This is progress. Most large IT departments are somewhere in this stage already.

The limitation is that each tool monitors its own perimeter. The IAM alert and the ERP exception and the GRC finding all sit in separate queues, reviewed by separate teams on separate schedules. The cross-system picture still doesn't exist. You have better data than you did at stage one, but the interpretation is still siloed.

Stage three: Integrated continuous monitoring. This is where the discipline we named earlier starts to apply. Controls from different systems get mapped to a common framework. An access change in the IAM system gets connected to the downstream SoD implications in the ERP. A configuration change in the production environment gets linked to the ITGC testing record for that system. Evidence from disparate sources gets normalized into a shared language.

At this stage, cross-system gaps become visible. You can see that two controls are each passing their own tests while the risk they're jointly designed to address is still present. You can identify when the timing between a control in one system and a dependent control in another creates a window of exposure. The picture starts to cohere.

Most organizations find this stage technically achievable but organizationally difficult. The data flows across system boundaries. It requires someone to own the cross-system view rather than just the within-system view. That ownership question is exactly why the Enterprise Risk Orchestration function we described earlier matters. Without someone whose job is the whole picture, integrated monitoring tends to stall at the point where it crosses a departmental boundary.

Stage four: Continuous assurance. At this stage, evidence is continuously produced, continuously monitored, and continuously mapped to risk assertions. A control isn't described as "tested and passed in Q3." It has a current status based on evidence from the last twenty-four hours, the last week, the last month. The question "is this control operating effectively?" can be answered right now, not in reference to the last testing cycle.

This is where assurance becomes genuinely real-time. Issues are caught as they arise. Remediation happens before problems compound. Audit preparation shifts from "collect evidence" to "export the record." The people responsible for financial reporting have an actual answer to the question "are our controls working?" rather than an answer that's twelve months stale.

Stage five: Predictive control intelligence. At the leading edge, the system doesn't just tell you what controls are doing now. It identifies patterns that suggest where controls are likely to degrade. A configuration that keeps drifting back to a non-compliant state. An access pattern that keeps approaching the boundary of a segregation-of-duty policy. A process timing sequence that repeatedly comes close to triggering an exception. The system surfaces these patterns before they become findings.

This is the direction the field is heading. I'll come back to what it requires technically when we reach the AI accelerant. For now, the important point is that stage five isn't a leap from stage one. It's the end of a progression, and you can start moving along that progression today regardless of where your environment currently sits.

The practical guidance is straightforward. Identify your two or three highest-risk process areas. Map the controls that address those risks across all the systems involved. Understand what evidence each of those controls currently produces and how often. Then ask what it would take to make that evidence flow continuously and get reviewed without waiting for an audit request. You'll find that in many cases the evidence is already flowing. The gap is in the routing, the normalization, and the review.

That's a smaller problem than rebuilding your control environment from scratch. And it's where continuous assurance actually starts.

MEASURING CONTROL CONFIDENCE, NOT CONTROL COUNTS

There's a number that appears in a lot of board presentations: the total count of documented controls. Companies with large, complex environments often run into the hundreds or even thousands of individual control points. The number tends to grow over time, as each new audit finding prompts the addition of a compensating control, each new regulation requires documentation of existing practices, and each new system brings its own control inventory.

A high control count is treated as evidence of a mature control environment. More controls means more protection.

I've never found that to be true in practice.

What I've found is that organizations with very high control counts often have among the worst visibility into whether any individual control is actually working. The count grew because people added controls faster than they retired old ones, because documenting a control is easier than deleting one, and because nobody wanted to explain to an auditor why a control was removed. The number went up. The clarity went down.

The better measure is control confidence.

Control confidence asks a different question. Rather than "how many controls do we have?" it asks "how certain are we that each control is operating as designed, right now?" The answer isn't a count. It's a measure built from the evidence that control has produced, the frequency of that evidence, the pattern of exceptions, and the trend over time.

Think about how this works in practice. You have a control that requires dual approval for all vendor payments above a specified threshold. The control is documented. It's been tested. It passed the last audit. Control count: one. Now ask a different set of questions. How many payments above that threshold went through last month? How many had two distinct approvers? Were any of those approvers the same person using two different credentials? Were any payments split to stay just below the threshold? Were the approvals contemporaneous with the transaction, or did they come hours or days later?

Those questions don't change the control count. They change your confidence in whether the control is actually doing what it says.

An organization with six hundred controls and low average confidence across that inventory is in a worse position than an organization with two hundred controls and high confidence in each one. The first organization has spent a lot of effort documenting things. The second organization has spent its effort on actually understanding whether its controls work.

Control confidence has several components, and it's worth being precise about what goes into it.

Evidence frequency is the first component. A control that produces evidence every day gives you much better confidence than a control that produces evidence once a quarter. Not because daily evidence is inherently more valid, but because it leaves much less time for something to go wrong between data points. If your segregation-of-duty control produces a report every morning, you'll catch a violation within twenty-four hours. If it runs quarterly, you might be sitting on ninety days of exposure.

Evidence completeness is the second. A control that tests samples has lower inherent confidence than a control that tests everything. Sampling is often necessary and entirely appropriate. But when the risk in a given process area is high enough, the cost of a full-population test is usually worth paying. Modern systems can run full-population checks on most transaction-level controls without significant overhead. The question is whether anyone has decided to use that capability.

Exception patterns are the third. A control with zero exceptions sounds reassuring. Sometimes it is. Sometimes it means the control isn't being triggered at all, or that exceptions are being resolved outside the system. The pattern of exceptions over time, including near-misses and corrected exceptions, tells you more than the presence or absence of violations in any given period.

Trend direction is the fourth. A control whose confidence score has been declining over three months even though it hasn't formally failed yet is giving you a signal. Configuration drift, staff turnover, process changes that outpace system updates: these show up in the trend before they show up in a formal exception. A control that was passing cleanly six months ago and is passing with increasing exception frequency today is a control that's heading toward a failure. The trend is the early warning.

Put these components together and you get something much more useful than a checkbox. You get a current, evidence-based view of how much confidence you should actually place in each control. And when you aggregate that across a control inventory, you get a genuine picture of whether the business is in control, not just a picture of how many controls you've documented.

This is the shift the periodic model can't make. When you test a control once a year, you get a binary outcome: it passed, or it didn't. You don't get frequency, completeness, pattern, or trend. You get a point-in-time verdict on a sample. Continuous assurance replaces that verdict with a living confidence score, updated by the evidence the control itself produces as it operates.

The CFO who asks "are our controls working?" deserves a better answer than "we passed SOX last year." They deserve a current answer, built from current evidence, that they can actually rely on when they sign the representation letter or brief the audit committee. That answer requires continuous assurance. It requires moving from counting controls to measuring confidence. And it requires treating evidence production as a core operational function, not an audit-time artifact.

MAKING THE TRANSITION

The case for continuous assurance is clear. The path to get there requires some honest assessment of where you're starting.

Most finance and risk teams that want to move toward continuous assurance run into the same three friction points.

The first is data access. Evidence sits in systems that weren't designed to share it. SAP produces logs. NetSuite produces logs. Your IAM platform produces logs. Getting those logs into a place where they can be cross-referenced requires integration work that often sits at the bottom of IT's backlog. The practical approach is to start with the systems where the data is already accessible and build outward. You don't need full integration on day one. You need one process area where you can demonstrate what continuous monitoring of evidence looks like, then use that proof point to move the integration work up the priority list.

The second is ownership. Continuous assurance requires someone to watch the evidence continuously. That's a different operating model than assigning a controls tester to audit preparation twice a year. The question of who owns daily review of control performance is an organizational design question, and the answer usually requires some reallocation of responsibilities between the controls team, internal audit, and the process owners who are accountable for the underlying transactions. The orchestration function discussed earlier is the right home for this capability, but building it requires explicit decisions about who does what.

The third is culture. Periodic assurance, for all its limitations, gives people a comfortable rhythm. Preparation, fieldwork, findings, remediation, done. Continuous assurance removes the comfortable gaps between events. Issues surface faster, which means they require responses faster. Some teams experience this as pressure rather than improvement. Managing that transition requires framing continuous assurance as a tool that makes the team's job easier over time, not a mechanism for catching people faster. The goal is fewer surprises at year-end, not more alarms on a daily basis.

None of these friction points are insurmountable. They're predictable, which means they're addressable. The organizations that have moved furthest along the maturity path started by picking one process, demonstrating a working model, and expanding from there. The sprawling, all-at-once transformation is the version that stalls. The focused, evidence-led version gains traction.

WHAT YOU CAN DO THIS QUARTER

The gap between periodic and continuous is real, but it closes in practical steps. Here are four that don't require a technology overhaul to begin.

Start with your three highest-risk controls. Identify the controls that address your largest financial exposure or your most common audit findings. Map every system that touches those controls. Then list what evidence each of those systems is producing today, how often, and who is currently reviewing it. That inventory alone will show you where the gaps are.

Establish a baseline for control confidence. For each of those three controls, ask the four questions: How frequently does this control produce evidence? Is the evidence complete or sample-based? What has the exception pattern looked like over the past six months? Is the trend improving or declining? You don't need a formal tool to do this the first time. A spreadsheet and a conversation with the system owners will get you most of the way there.

Move one control to daily evidence review. Pick the highest-risk item from your list and change the monitoring frequency from quarterly or annual to daily. Set up an automated alert for exceptions. Assign someone to review the output each morning. Run it for ninety days and document what you find. That ninety-day record will be more valuable than any control testing documentation you could produce, and it will build the internal case for expanding the model.

Define what "audit ready" means as a permanent state. Write down what evidence your external auditors would need to close fieldwork on your three highest-risk areas. Then ask how much of that evidence could be automatically produced and maintained on a rolling basis, rather than collected in a sprint before fieldwork begins. The gap between the two lists is your implementation roadmap.

The goal isn't to eliminate the annual audit. External assurance has a legitimate role, and that won't change. The goal is to stop treating the annual audit as the primary signal of whether controls are working. That signal should be continuous, current, and internally owned. The audit becomes a verification of what you already know, not a discovery process that tells you things you should have known months ago.

That's the operating model this chapter is about. Not auditing faster, but building an environment where the evidence of control performance is always present, always current, and always owned by someone whose job is to act on it.

Chapter 12
The AI Accelerant

For years, the fragmentation problem described in this book was serious but manageable. Systems sat in silos. Tools produced disconnected outputs. Risk hid in the gaps. But the pace of the whole thing was human. Someone had to click approve. Someone had to run the batch. A person was somewhere in the chain.

That's changing fast.

For many enterprises, it's already changed. The ERP doesn't wait for you. The workflow doesn't pause while you think. Automated payment runs, automated vendor onboarding, automated journal posting, automated reconciliation, automated controls testing. Modern ERP environments can process thousands of financial transactions per hour with minimal human intervention.

The technology isn't the problem. Automation is genuinely good. It reduces manual error, speeds settlement, cuts cost, and scales what would otherwise require armies of people. Nobody serious is arguing to roll it back.

The problem is that our assurance model didn't keep pace.

AUTOMATION AT SCALE

When I started in this field, "automation" in an enterprise finance context mostly meant batch jobs overnight. You'd run the AP payment batch at 2am. Someone would review the output in the morning. There was a natural lag built into the process, and humans sat at most of the checkpoints.

That world is largely gone.

Today, a well-configured S/4HANA environment with connected procurement and supplier networks can process a purchase order, match it to a goods receipt, approve the invoice, release the payment, and post the accounting entry in a sequence that takes minutes. Across a mid-sized enterprise, this might happen hundreds of times a day. Across a large one, thousands. The same pattern plays out in order-to-cash, in payroll, in intercompany eliminations.

The efficiency gains are real and significant. That's why every major enterprise is pursuing this path.

But here's what doesn't scale at the same rate: human judgment in the control layer.

The access control you defined two years ago still governs who can approve payments today. The SoD rules configured in your GRC platform were set up during the last implementation. The monitoring thresholds in your SIEM were established when transaction volume was a fraction of what it is now. The control logic was set by people, at a point in time, based on what the environment looked like then.

Meanwhile, the environment kept moving.

Roles changed. Processes got re-engineered. New modules went live. Integrations were added. Custom code was modified. All of it happened while the underlying control configuration mostly held still. We covered how fragmentation accumulates quietly in earlier chapters. But automation makes the consequences arrive faster and at greater scale.

A manual process that breaks a control runs at human speed. An automated process that breaks a control runs at machine speed.

That's the core tension. And it's accelerating.

THE QUESTION ALMOST NOBODY IS ASKING

There's a version of this conversation already happening in most boardrooms and CFO offices. "Are we keeping up with AI?" "What are we doing with automation?" "How do we get more out of our ERP investment?" These are good questions and they deserve serious answers.

But there's a question that almost nobody is asking alongside them: when we automate a process, who is accountable for whether the controls on that process still work?

Automation doesn't transfer accountability. It obscures it.

When a human approves a vendor payment, you can identify who approved it, when they did it, what they saw when they made the decision, and whether they had the authority to make it. The control is visible. When an automated workflow approves the same payment based on matching logic, tolerance bands, and pre-approved supplier relationships, the answers to all of those questions live in configuration files, vendor tables, and workflow rules that most finance and audit teams haven't looked at since go-live.

The control exists. Technically. But whether it's operating as intended, whether it still reflects current policy, whether it's been drifted by a system change or a workaround, whether it would catch the thing it was designed to catch, those questions don't have clean answers.

Most organizations assume the automated control is working because the system is running. That's a comfortable assumption. It's also usually wrong in at least a few places at any given time.

I've spent close to thirty years watching enterprises implement controls, and the pattern is consistent: the most dangerous gaps aren't in areas where people know they have weak controls. They're in areas where people believe they have strong ones. Automated controls are especially susceptible to this. They were designed carefully, tested at go-live, ran without incident for a year or two, and so they get reclassified in everyone's mental model as "handled." Meanwhile, the configuration that governs them sits untouched while everything around it changes.

Now layer AI on top of this.

The current wave of AI deployment in enterprise environments isn't mostly about chatbots and content generation. It's about agents: AI systems that take action on behalf of the business. Workflow automation agents that route approvals. Reconciliation agents that clear matching items and flag exceptions. Procurement agents that select and qualify vendors. Finance agents that propose journal entries. These agents don't just surface information for a human to act on. They act.

And when they act, the question of whether their actions fall within the organization's intended control boundaries is almost entirely a function of how well those boundaries were defined and how recently they were validated.

Most weren't defined with AI agents in mind. Because most were defined before AI agents existed.

CONTROL DRIFT AT MACHINE SPEED

"Control drift" describes what happens when a control that was correctly designed and properly implemented gradually moves out of alignment with the environment it was meant to govern. It's not fraud. Nobody made a bad decision. The control was right when it was built. The environment just moved around it.

In a manual environment, control drift is a slow problem. People notice when something feels off. Approvals that used to take two steps now take one, and someone eventually asks why. A workflow that was supposed to require a manager sign-off starts routing around it, and someone eventually notices the audit trail gap.

In a high-automation environment, control drift is fast and quiet. The system doesn't feel off. It runs exactly as configured. The problem is that the configuration no longer reflects the intent.

Consider the typical life cycle of an automated payment release control. Someone in the finance team designs the logic: invoices below a certain amount with a matching PO and goods receipt can auto-release; everything above goes to workflow approval. Good control. Tested. Signed off.

Eighteen months later, the tolerance band gets widened during a period of high invoice volume because the approval queue was creating cash flow delays. Someone makes that change in a configuration table. It goes through change management, technically. But the risk implications aren't fully assessed. The new band wasn't remapped to the original control design. Nobody checked whether this specific change invalidated the assurance the control was supposed to provide.

Another year passes. The company adds a new business unit with different supplier relationships. The vendor master gets expanded. Some of the new vendors were onboarded under slightly different approval criteria because the standard process was too slow for the acquisition timeline.

Now your auto-release logic is operating against a vendor population it was never designed for, at a tolerance level that was bumped up under operational pressure, in a configuration that three people changed over two years and none of them saw the full picture.

The control is still "in place." Every audit checklist will mark it green. But the control you have now is materially different from the control you designed. And the difference lives in the gaps between the change management system, the vendor master governance process, and the payment configuration table, none of which talk to each other.

This is control drift. AI agents make it structurally worse.

Every AI agent that takes action in your environment is operating against rules, constraints, and data that someone defined at a point in time. As the environment changes around those rules, the agent keeps acting according to the original configuration. It doesn't notice the drift. It doesn't flag the gap. It just executes.

An agent authorized to select vendors from an approved list will keep selecting from that list even if the list hasn't been reviewed in fourteen months. An agent authorized to clear reconciling items below a threshold will keep clearing them even if the account's risk profile has changed since the threshold was set. An agent authorized to route journal entries for review above a materiality level will keep using that level even if the company's size has grown and the level is now too low to catch what matters.

In each case, the agent is doing exactly what it was told to do. The problem is that what it was told to do is no longer aligned with what the business actually needs it to do.

This is the AI assurance problem. Most enterprises don't yet have a coherent answer to it.

THE ASSURANCE LAYER THE AI ERA DEMANDS

There's a temptation at this point in the conversation to propose a new tool. More monitoring. Another layer of software. A dashboard for the dashboards. I understand why people go there, because something clearly needs to change. But adding another point solution to a fragmented stack doesn't solve the problem I've just described. It adds another silo.

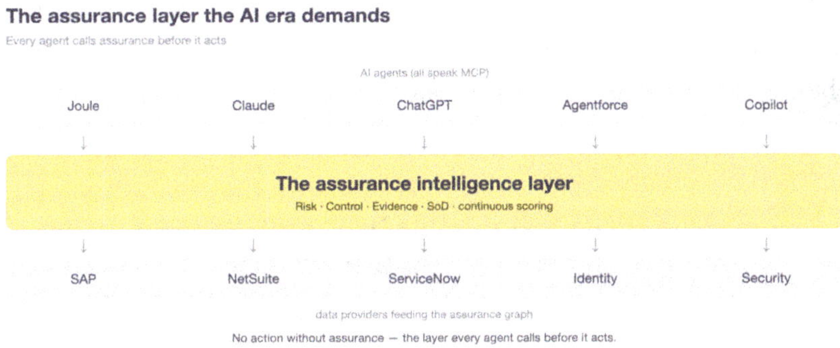

What the AI era actually demands is coherence: the ability to see across systems, agents, configurations, and changes, and to know whether the control layer is actually doing what it was designed to do.

The four planes of assurance apply here, but with new urgency.

Intent and authority. When an AI agent takes an action, can you trace that action to an explicit authorization? Was the agent acting within the scope it was given, at the privilege level it was assigned, under governance rules that a human being consciously approved? If you can't answer that in detail, you have an assurance gap. This is the same SoD problem that has existed in SAP environments for decades, extended to cover automated actors alongside human ones.

Timing. Is the assurance you have about an agent's controls current? When were the rules it operates under last validated against current business intent? In a continuous assurance model, this question gets answered on a rolling basis. In a periodic model, the answer might be "fourteen months ago," which is not good enough when the agent is acting thousands of times a week.

Evidence. When an AI agent acts, does that action generate a record that can be audited? Can you reconstruct what the agent did, why it did it based on the rules it was operating under, and whether those rules were appropriate at the time? Automated actions that don't produce auditable evidence are a material weakness waiting to be discovered.

Orchestration. Is there a layer in your environment that aggregates the control signals from AI agents alongside the control signals from your human-operated systems, and surfaces a coherent view of whether the whole thing is in control? If each agent's assurance is managed in isolation, you're back to the silo problem, just with faster actors.

The organizations that get this right won't necessarily have the most sophisticated AI. They'll have the most coherent assurance. The agents will be able to act quickly precisely because there's a control layer that can validate their actions in near-real time against a consistent picture of intent, authority, evidence, and timeliness.

Think of it this way. A commercial pilot can fly faster, in worse weather, with less visibility than a private pilot. Not because of better instincts, but because of better instruments and a more robust set of procedures that keeps them inside known safe parameters at every step. Speed becomes possible when the assurance layer is strong enough to support it.

The same logic applies to enterprise automation. The question for the CFO or CRO reviewing an AI deployment isn't just "what can this agent do?" It's "what does the assurance layer look like that governs what this agent does?" If there isn't a clear, confident answer to the second question, the first question is answering itself in a way the board probably won't enjoy reading about later.

It's worth addressing where the largest ERP vendor is taking its own controls, because a reader inside an SAP shop will have heard the announcements. SAP is moving its governance and risk products onto a single, in-memory platform and embedding generative AI into control monitoring and risk indicators, with the platform able to test entire transaction populations continuously rather than sampling, and an early initiative to put AI agents to work on control and compliance tasks.

This is a genuine advance, and it points in the same direction this book argues: from retrospective sampling toward real-time, fuller-coverage assurance. It would be a mistake to treat it as either irrelevant or as the whole answer. It is a tailwind — confirmation that the destination is real — and a powerful new source of signal for the orchestration layer to read.

What it doesn't do is close the coherence gap, because that gap was never only inside SAP. A faster, smarter control plane that sees the SAP estate is still one instrument in the orchestra. Most enterprises run more than one ERP and a wide field of non-SAP systems that carry financial risk, and no single vendor's platform — however modern — reads across all of them. The orchestration discipline sits above any one vendor's tools, including the best of them, precisely so that it can.

TIMELY, NOT ALARMIST

I want to be deliberate about the tone here, because the AI-and-risk conversation has a habit of sliding toward one of two unhelpful places. The first is panic: AI is ungovernable, controls are useless, the enterprise is flying blind. The second is dismissal: these are just normal technology risks, we've always had automation, nothing fundamentally new here.

Both are wrong, and both make the work harder.

The panic framing causes organizations to either over-respond with governance theater, layers of approvals and sign-offs that don't actually address the underlying assurance gap, or to freeze on AI adoption entirely while competitors move forward. Neither outcome serves the business.

The dismissal framing causes organizations to assume that existing controls extend naturally to AI agents when they don't. The discovery usually comes later, in an audit finding or a material weakness disclosure or a board-level incident conversation, that the assumption was wrong.

The honest position is this: the risks are real, they're growing, and they're addressable. The enterprises that address them well will get the full benefit of automation. The ones that don't will find that their automated environments generate the same fragmentation problem they already had, just at greater speed and scale.

What I've seen over nearly thirty years is that enterprises that handle technology transitions well do one thing consistently. They don't ask "how do we control this new thing" in isolation. They ask "how does this new thing fit into the assurance picture we already have?" That's a better question because it keeps you from building another silo.

An AI agent deployed into an environment with coherent, continuous assurance, where intent and authority are clearly defined, where evidence is captured automatically, where the control configuration is monitored for drift, where timing is tracked and validated, that agent is governable. It's a capability with appropriate guardrails, not a risk multiplier.

An AI agent deployed into the fragmented environment most enterprises actually have, where controls are siloed, evidence is episodic, authority boundaries are fuzzy, and nobody has a cross-system view of whether anything is working, that agent will find the gaps just as surely as the automated payment processes, the manual journal entries, and the SoD conflicts did. Faster, and at greater volume.

THREE MOVES THAT MATTER NOW

You don't need to wait for a full-scale AI deployment to start preparing your assurance layer. These three moves are practical regardless of where you are on the automation curve.

Audit your automation inventory, not your AI inventory. Most organizations have dozens of automated processes running in their ERP and adjacent systems that have never been systematically reviewed for control drift. Before you add AI agents to the picture, you need to know the state of the automated processes already running. Map them. Find the ones where the original control design hasn't been reviewed since implementation. That review will tell you whether your current assurance model can support the automation you already have, which is the prerequisite for the automation you're planning.

Define authorized action before any agent goes live, and build the evidence trail into the deployment. Every AI agent your organization deploys should have explicit, documented authorization boundaries: what it can do, what it cannot do, under what conditions its authority applies, and what creates an exception that requires human review. That documentation isn't governance overhead. It's the baseline against which you'll detect drift. Without a baseline, you can't measure the gap.

Put AI assurance on the same agenda as AI capability. The conversation about what AI agents will be able to do is already happening at the board and executive level in most organizations. The conversation about what the assurance layer looks like needs to be in the same room, not a separate risk committee item that comes along six months later. The same room, the same week, ideally the same conversation.

The organizations that will look back on this period and feel good about their decisions are the ones that treated the AI opportunity and the assurance question as inseparable. Capability and control, both at the same time.

Speed with coherence. Automation with accountability. The ability to deploy AI agents with confidence because the assurance layer underneath them is actually working.

That's the promise of Enterprise Risk Orchestration applied to the AI era. Next, we look at what the new standard of governance actually requires, and what boards and CFOs should now be asking to know whether their enterprise is genuinely in control.

Chapter 13
The New Standard

The conversation in most boardrooms today still runs on two words: "we're compliant."

Those two words close the topic. They end the question. They move the meeting forward. And for the past two decades, they've been accepted as a reasonable answer to a question that deserves a much harder one.

The harder question is: "But are we actually in control?"

Compliant and in control are not the same thing. By now, if you've read through this book, that distinction should feel obvious. Controls can exist on paper while failing in practice. Systems can pass every audit test while running at speed through gaps nobody mapped. Organizations can check every box, clear every review, and still be operationally exposed in ways that won't surface until something breaks.

The discipline of Enterprise Risk Orchestration exists to close that gap. The four planes give you a framework. The enterprise landscape gives you a map. Continuous assurance gives you a live signal instead of a rearview mirror. And the AI accelerant makes the whole thing urgent in a way that periodic compliance cycles simply can't match.

But frameworks and disciplines only change things when someone decides to hold a different standard. That decision usually starts in three places: the boardroom, the CFO's office, and the internal case that gets made to leadership by the people who live inside the problem every day. This chapter is about each of those three.

WHAT BOARDS SHOULD NOW BE ASKING

Audit committees have gotten better over the past twenty years. SOX forced the issue. The financial crisis added more pressure. Cyber incidents, regulatory actions, and some very public restatements made boards take the internal controls conversation more seriously than they once did.

But "more seriously" still mostly means asking the same questions with more intensity. "Is our SOX program in good shape?" "Did we have any material weaknesses?" "What did external audit find?" Those are reasonable questions. They're just the wrong ones for the operating environment we're actually in.

A board that only asks about SOX attestation is a board governing a 2004 risk model in a 2026 enterprise. The problem isn't that they don't care. The problem is that nobody gave them better questions.

Here are the questions that better-informed boards are starting to ask, and that every audit committee should have in front of them.

"What percentage of our controls are tested continuously versus periodically?"

This question immediately reveals whether the organization is operating with a live signal or a historical one. Most enterprises test the overwhelming majority of their controls on a periodic basis: quarterly, semi-annually, annually. A handful of automated controls get tested continuously. The rest get tested in a window that conveniently aligns with audit season. If the answer to this question is "less than 20% continuously," the board is essentially governing with data that's always at least 90 days old.

"Where do our systems not talk to each other, and what risk lives in that gap?"

The integration layer is the blind spot we covered in the chapter on the enterprise landscape. Most boards have never seen a map of it. They've seen functional org charts and legal entity diagrams and technology spend reports. Almost none of them have seen a diagram showing where their core ERP connects to their middleware, where that middleware connects to their cloud applications, and which of those handoffs has no control, no monitoring, and no owner. That diagram exists in most enterprises if you build it. Boards should be asking for it.

"When did we last test whether an exception actually triggered a human review?"

Controls that require human review are only as good as the process that routes exceptions to actual humans who read them, evaluate them, and act. Most organizations can tell you the exception threshold. Many can tell you how many exceptions fired last quarter. Very few can tell you how many of those exceptions resulted in a documented human decision within the required window. That last number is the one that matters, and it's almost never tracked.

"What's our irreversibility point for the highest-risk transactions in our system?"

This is the question from the four planes that the AI era makes unavoidable. For some transactions, you have hours. For others, you have minutes. For wire transfers in certain configurations, you have seconds. The board should know which processes in their enterprise carry the shortest window between execution and irreversibility, and they should know whether controls sit upstream or downstream of that window. Downstream controls, no matter how well designed, can't undo a transaction that's already settled.

These questions don't require the board to become technical. They require the board to hold a higher standard than attestation. And they signal to management that "we passed the audit" is no longer a sufficient answer to "are we in control."

There's a practical consequence to this shift. When boards ask better questions, management prepares better answers. When management prepares better answers, they need better evidence. When they need better evidence, they build better systems for producing it. The standard sets the direction. The direction determines the investment.

If the board keeps accepting attestation, attestation is what they'll get. If they start asking for evidence, they'll start getting it.

THE CFO'S MOMENT

The CFO sits at an unusual intersection in this conversation. They own the financial reporting process, which makes SOX and audit results very much their problem. They own the ERP environment, or at least the budget for it. They're accountable to the audit committee for the accuracy of what gets reported. And they're increasingly expected to give a real-time view of the business, not a quarter-end view.

That combination makes the CFO both the most exposed person in the organization when assurance gaps surface and the most positioned person to close them.

The CFO who wants to lead this conversation has a specific opportunity right now, before the AI wave fully hits and before regulators catch up with what AI-driven automation actually requires from a controls standpoint. The window to build the assurance infrastructure proactively is open. It doesn't stay open indefinitely.

What does leading the conversation look like in practice?

Start with the close. Every CFO knows the financial close process carries more risk than almost anything else the organization does. Revenue recognition, journal entries, reconciliations, intercompany eliminations: these are the processes where errors compound, where manual interventions multiply, and where the audit always seems to find something. The close is also the process where the symptoms of assurance gaps show up most visibly. If the close takes ten days when it should take five, the extra five days are usually spent chasing evidence that should have been captured in real time throughout the month.

A CFO who maps the close process against the four planes will find something instructive. On the intent and authority plane: who is actually authorized to post those late journal entries, and does the system enforce that authorization or just record it? On the timing plane: which reconciliations are happening after the fact rather than as the underlying transactions occur? On the evidence plane: what documentation exists to show that each step happened, in the right sequence, with the right approvals, and not just that the output looks right? On the orchestration plane: which systems contributed to the close, and do any of those handoffs have no control attached?

That exercise takes a few hours with the right team. What it produces is a map of where close risk actually lives, not a list of what got tested, but a picture of what the process actually depends on to work correctly. That map is worth more than most audit reports, because it shows the structure of the risk, not just the symptoms that showed up last quarter.

The CFO can also change the conversation with the external auditors. Most CFO-auditor relationships center on agreeing on materiality thresholds, resolving open items, and managing the timing of the audit. A CFO who shows up with a continuous evidence package changes the dynamic. The auditors can query evidence directly. They can see that controls operated throughout the period, not just at the point of testing. They spend less time reconstructing the past and more time validating the picture the organization has already built. Audits get shorter. Findings get fewer. And the CFO stops dreading the annual scramble to produce documentation for things that happened eight months ago.

One other point that CFOs sometimes miss: the assurance infrastructure they build for financial controls is the same infrastructure that supports every other enterprise risk conversation. When the CEO wants to know whether the procurement controls held during a high-growth period, the CFO with an evidence engine can answer that question in hours. When the audit committee wants to know whether access controls changed after a system upgrade, the CFO with a continuous monitoring platform can pull the record. When external counsel needs documentation for a regulatory inquiry, the evidence is already assembled. The CFO who builds the assurance layer stops being the person who gets asked hard questions they can't immediately answer.

That's a different relationship with risk than most CFOs have right now. It's also an achievable one.

MAKING THE INTERNAL CASE

If you're a CAE, a risk officer, a controls leader, or an ERP security owner reading this, you probably already know the discipline I've been describing exists. You've been living in the gap for years. You know the tools don't talk to each other. You know the annual audit cycle can't keep pace with how fast the business moves. You know the integration layer carries risk that nobody's mapped. You've been saying versions of this in meetings for longer than you'd like to admit.

The problem is usually not knowledge. The problem is the internal case.

Getting organizational attention and budget for assurance infrastructure is hard because the thing you're building is, at its core, invisible when it works. The controls are operating, the evidence is being captured, the exceptions are getting reviewed. Nothing dramatic happened. That's the product. And in most organizations, the budget conversation rewards things that are visible and urgent, not things that quietly prevented something bad from happening.

So the case has to be built differently.

Start with a specific, bounded failure. Don't lead with the abstract argument about fragmentation (even though you're right). Lead with one real example from your own environment where a control existed, passed its last test, and still failed to catch something that cost the organization money, time, or audit findings. You don't need it to be a catastrophe. A single instance where the annual audit found a material issue that a continuous monitor would have surfaced months earlier is a concrete anchor for the conversation. Specificity makes the problem real in a way that frameworks don't.

Then connect it to a number. The cost of assurance gaps isn't always direct. Sometimes it's the cost of the audit itself: the hours spent reconstructing evidence, the fees paid to external auditors for extended fieldwork, the management time consumed by remediation. Sometimes it's the cost of the finding: the restatement, the regulatory inquiry, the reputational impact with the audit committee. Sometimes it's the opportunity cost of the close taking three weeks instead of one. Whatever the number is in your organization, attach it to the example. "Last year, we spent an estimated X in additional audit fees because we couldn't produce evidence that controls operated during Q2." That's a number leadership can work with.

Then show the maturity path, not the full solution. One of the most common mistakes in making the internal case for assurance investment is presenting the complete vision and letting leadership sticker-shock themselves out of the conversation. The orchestration layer, the continuous monitoring engine, the cross-system evidence graph: all of it together looks expensive and complex and easy to defer. Instead, show the first step. What's the highest-risk gap in your current environment, based on the landscape map you built earlier? What's the single integration that carries the most unmonitored risk? What's the control category where you have the least continuous coverage and the highest frequency of audit findings? Start there. One plane. One gap. One measurable improvement.

The CAEs and risk leaders I've seen succeed with this conversation share a common quality: they speak the language of the person they're pitching to. With the CFO, they lead with close risk, audit cost, and evidence quality. With the CIO, they lead with the integration layer, the system handoffs, and the monitoring gaps in the technical architecture. With the CEO or the audit committee, they lead with the question of what happened between the last audit and today, and how confident the organization is in the answer. The underlying argument is the same. The framing shifts.

One thing to avoid in this conversation: selling the technology before you've sold the problem. The moment you lead with a tool or a platform, you've moved the conversation from "are we exposed" to "is this the right product," and you've made it easy for the conversation to stall on procurement timelines, IT review processes, and budget cycles. Lead with the exposure. Let the technology follow from the need.

FROM ATTESTATION TO EVIDENCE

The shift this book has been building toward isn't primarily technical. It's about knowledge — what you can actually know about your controls, and prove.

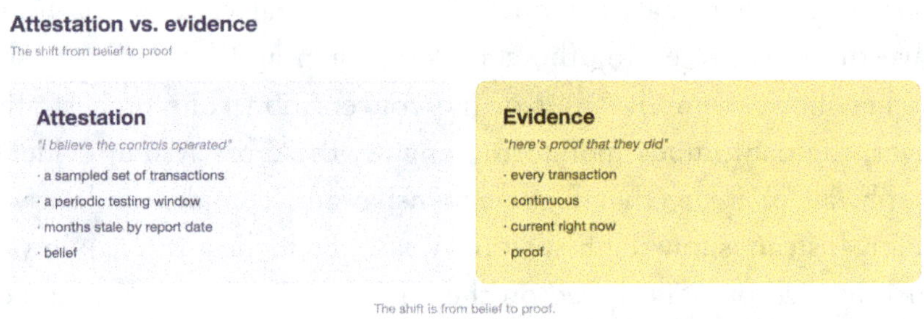

Attestation says: "I believe the controls are operating." Evidence says: "Here's proof that they did."

Attestation has served the compliance industry reasonably well since SOX passed in 2002. It created accountability where there was none. It forced management to actually think about controls rather than just assume they worked. It gave auditors a structured basis for evaluating the design and operating effectiveness of ICFR. Those were real improvements from a pre-SOX world.

But attestation has a ceiling, and we've been bumping against it for years. You can only attest to what you observed. You can only observe what your testing covered. Your testing covers a sample. The sample is drawn from a period. The period ended months before the report was issued. And the report lands on the audit committee's table and says "we believe our controls are effective" based on observations that are, in many cases, a year old by the time anyone reads them.

Evidence works differently. Evidence is produced continuously as controls operate. It's captured in the systems that execute the transactions, in the platforms that manage access, in the tools that monitor exceptions. It doesn't depend on a testing window. It doesn't compress a year of activity into a sample. It shows, record by record, whether the control operated in the way it was designed to operate, at the time the transaction happened, with the authorization that was in effect, and with the outcome that was expected.

The difference in confidence between those two positions is not small.

An organization running on attestation can tell you that their controls were designed effectively and appeared to operate effectively based on a sample of transactions tested during a specific period. An organization running on evidence can tell you that their controls operated on this transaction, and this one, and the 847,000 others that processed last quarter, and here are the 23 exceptions that fired, here are the 23 reviews that happened, and here are the 21 that cleared and the 2 that escalated to management action.

That second organization isn't just more defensible in an audit. It's genuinely safer. The controls are either working or they're not, and the organization knows which, in real time, rather than finding out six months later when someone asks about a quarter that's already closed.

Getting from attestation to evidence is a journey, not a switch. Most organizations will spend several years building the infrastructure, the processes, and the institutional habits that continuous assurance requires. The maturity model from the chapter on continuous assurance gives you the path. The four planes give you the architecture. The landscape mapping we walked through gives you the starting point.

But the direction should be clear. The new standard is evidence, not attestation. The new question is "show me" not "do you believe." The new conversation in the boardroom isn't "did we pass" but "what does the evidence say."

Boards that adopt that standard will hold management to it. Management that adopts it will demand it from their teams. Teams that operate under it will build the systems that produce it. And auditors, regulators, and investors who start seeing it will begin expecting it from peers who don't have it yet.

That's how new standards become standard.

WHAT TO DO MONDAY

You don't need a transformation program to start. You need a decision about where to look first.

Here's a practical sequence for the first thirty days.

Pull your last three audit reports and highlight every finding that involved a control that existed on paper but failed in execution. Don't analyze them yet. Just count them. That number is your baseline exposure rate under the current model.

Map one high-risk process end to end across every system that touches it. Pick the financial close, or the procure-to-pay cycle, or the order-to-cash flow. List every system that handles a step. For each system handoff, ask: is there a control at this point? Is it tested continuously or periodically? Who owns it? That exercise will surface two or three gaps you didn't know were gaps.

Take one finding from the audit report and work backward to the plane it failed on. Was the authorization intent not enforced at the system level? Was the control designed to operate in real time but only tested annually? Was there no evidence that a human review actually happened, even though the control required one? Locating the plane tells you what kind of fix is needed.

Bring one specific, costed exposure to your CFO or audit committee. The gap you found in the landscape map, the control that failed in a way that cost something, the evidence you can't produce for a question you're being asked. One specific thing, with a number attached. See whether it changes the conversation.

Those four steps don't require budget approval or technology procurement. They require attention and a willingness to look at the enterprise the way it actually operates rather than the way the control documentation says it should.

The organizations that move first on this won't just have a better audit story. They'll have a genuine operating advantage: faster closes, faster audits, faster answers to the questions that boards and regulators are increasingly asking. They'll know what's actually happening in their systems, not what happened six months ago in a sample.

That's the new standard. The question now is who sets it first.

Chapter 14
The Category Is Named

Thirty years ago, I walked into my first major ERP enterprise engagement and found myself staring at something that not only didn't have a market name yet but wasn't on the radar of the company or the implementation team. The company had some controls on paper. They had auditors, systems, a few policies, and binders full of documented procedures. A controls team was actively documenting configurable controls in their system and so they had passed almost every review that mattered. And yet they faced a financial reporting problem that none of those controls had caught, because early on application controls and operational controls weren't designed to talk to each other.

I didn't know what to call it then. I knew about business operational processes and financial accounting controls and as I learned more about the implementation of these ERPs, something was missing from the picture. The group I worked with had one focus: ensuring controls were implemented. But the group was fairly new, application controls were new, and there were no tools on the market to assist.

That picture stayed with me through every engagement that followed. Through years at PwC, Deloitte, Accenture, SAP, KPMG, and others. Through hundreds of client environments, across industries, continents, and every major version of SAP that's been released in the last three decades. Even with all the new solutions to address controls, the picture kept showing up. New clients, new systems, new auditors, same problem. Excellent tools. Fragmented outputs. Risk hiding in the gaps.

The industry kept solving pieces of it. Better GRC platforms. Smarter identity tools. Continuous monitoring modules. Richer audit software. Each solution got sharper, more capable, more expensive. The fragmentation got worse anyway, because adding tools without adding coherence just creates more gap surface area.

Nobody named the whole problem. So nobody solved it.

This book is my attempt to do the naming.

THIRTY YEARS TO THIS MOMENT

The discipline I've described in these pages didn't arrive fully formed. It came from watching the same failure modes replay across different organizations, different regulatory regimes, different technology stacks. It came from asking, after every engagement, why controls that looked solid on paper kept failing in practice.

The answer, almost every time, was the same one we opened with. The enterprise wasn't under-controlled. It was under-observed. The signals were there. They lived in the GRC platform, in the access reports, in the SAP transport logs, in the audit trail, in the reconciliation exceptions. But they lived separately, in different systems, in different formats, reviewed by different teams at different intervals. No one was reading them together.

That's not a technology failure. It's a discipline failure. Specifically, it's the failure to treat cross-system coherence as a function, not as an accident.

When I started mapping the four planes of assurance, I wasn't inventing a framework for a book. I was documenting what I'd seen work when organizations actually got ahead of their risk. The enterprises that stayed in front weren't the ones with the most tools. They were the ones that had someone, or some function, asking: do our controls agree with each other? Does what the system is doing match what leadership authorized? Does the timing of our evidence match the timing of our risk? Is anyone actually reading the whole picture?

Asking those four questions consistently is Enterprise Risk Orchestration. The tools you already have can answer them. The gap is the asking.

Thirty years gives you a certain kind of perspective. You stop being surprised by the failures, and you start being able to predict them. Not because organizations are careless, but because the failure modes are structural. When you have eight systems producing evidence that nobody cross-references, gaps are guaranteed, regardless of how good each system is. That's not a management problem. It's an architectural one, and it has an architectural solution.

The solution is the orchestration layer: the function, the process, and the discipline that takes what your tools already produce and reads it as a single coherent picture.

That's what this book has been building toward. And the reason it took a full book to get here is that the solution can't be understood without understanding the problem at depth. Most organizations have normalized the fragmentation so thoroughly that they don't see it as a problem anymore. They see it as the cost of complexity. They've accepted "compliant yet exposed" as an inevitable condition of running a large enterprise.

It isn't inevitable. It's a choice, and it's one you can now make differently.

WHAT YOU CAN DO MONDAY

The field exists whether or not you have a formal Enterprise Risk Orchestration function yet. The concepts in this book work at any maturity level, in any organization that runs ERP. You don't need to rebuild your GRC stack. You don't need new tools. You need to start asking different questions with the evidence you already have.

Here's what that looks like in practice.

Start with the landscape map. If you don't have a current, accurate picture of every system in your control environment and how they connect, that's your first action. Not a PowerPoint. A working reference that your team actually uses. Who owns what system? What evidence does each one produce? What goes into each one to make that evidence reliable? The chapter on seeing the whole picture gave you the structure for that map. The question now is whether it exists in your organization.

If it doesn't, make it. Even a rough version, built in a week with your most knowledgeable people, is more useful than the implied map everyone's been working from. The implied map is where the gaps hide.

Run the four-plane check on your highest-risk process. Pick one. Revenue recognition, procure-to-pay, payroll, financial close, whatever keeps you up at night. Work through it on each plane. Do the controls operating in that process match current leadership intent and authorization? Is the timing of your control testing aligned with when risk actually occurs, or are you testing quarterly against daily transactions? Is your evidence complete and current, or are there gaps between when the system logs something and when anyone reviews it? And are the controls in that process actually talking to each other, or are they operating in separate silos owned by separate teams?

You'll find something. Everyone does. That's not a failure; it's the whole point. Finding a gap before an auditor or regulator does is the entire value of the exercise.

Name one gap at the right level. The language you use matters. When you take a finding to your CFO or audit committee, the framing determines how it gets received. "We have a deficiency in our SoD controls" lands differently than "our finance system is approving transactions that our authorization policy doesn't allow, and we have no real-time visibility when that happens." The second framing is accurate. It's also specific enough to act on and significant enough to get attention.

The gap probably already exists in your data. What's usually missing is the translation from technical finding to business risk statement. That translation is part of the orchestration discipline.

Assign someone the cross-system question. Somebody in your organization needs to own the question of whether your control environment is coherent. This doesn't have to be a new headcount. It might be a reconfigured mandate for your CAE, or a new accountability for your Controller, or a workstream inside your existing risk function. What it can't be is nobody's job, which is what it is in most organizations right now.

The signal that nobody owns it: when you ask "who reviews control outputs across all our major systems together," you get silence, or a committee name that meets twice a year, or a finger pointed at a tool that nobody actively interprets. If that's where your organization is, you know where to start.

Set a baseline for control confidence. The chapter on continuous assurance covered this in detail, but the practical action is simple. For your five most critical controls, can you state today, with evidence, that they're operating? Not "we tested them in Q3" and not "the system is configured correctly." Operating, right now, as designed, producing the expected outcome.

If you can answer that for five controls, do it for ten. Then for your full SOX scope. Then for the broader control environment. The discipline of measuring confidence rather than counting controls is what separates a real-time assurance posture from a periodic audit posture. You build that discipline one control at a time.

Have the AI conversation before it becomes urgent. The chapter on the AI accelerant made the case for why this is time-sensitive. If your organization is using, piloting, or planning to deploy AI agents in any financial or operational process, the question of how those agents will be governed isn't a future problem. The agents will make decisions. They'll take actions. They'll do so at a speed and volume that no human review process can keep up with. The assurance layer needs to exist before the agents do, not as a retrofit.

You don't need to solve the whole AI governance question this week. You do need to start the conversation with your CIO or CTO about what the control framework looks like for AI-assisted processes. The organizations that get ahead of this will set the standard. The ones that don't will spend several years explaining to auditors and boards why their AI-enabled processes were operating without adequate oversight.

THE FIELD WE ARE BUILDING

Naming a discipline matters. Not as a branding exercise, but because language shapes attention. When something has a name, people can organize around it, hire for it, measure it, build careers in it, and hold each other accountable to it.

"Governance, risk, and compliance" became a category because the industry named it. Once it was named, it got job titles, certification programs, conference tracks, technology platforms, analyst coverage, and board-level attention. The category created the profession.

Enterprise Risk Orchestration needs the same treatment. The problem it addresses has existed for decades. The tools that feed it are already deployed in most large enterprises. The cost of not doing it is well documented, in every material weakness disclosure, every "compliant yet exposed" outcome, every breach that followed a clean audit opinion. What's been missing is the name, the framework, and the community.

This book is a contribution to that community. But it's one contribution, and the field will only develop if other practitioners, risk leaders, auditors, and ERP professionals start using the language, applying the frameworks, and sharing what they find.

If you've worked through these chapters and recognized your organization in the problems described, you're already part of the field. You've been practicing a version of this discipline informally, probably for years. What I'm asking you to do is make it formal. Give it a name. Put it in your team's mandate. Talk about it in your next audit committee presentation.

The academics and the standards bodies will catch up. They always do, a few years after practitioners have already moved. COSO, ISO, ISACA, and every major assurance framework will eventually incorporate the language of continuous, evidence-based, cross-system assurance. The concepts are too important and the timing is too clear for them not to. But the practitioners get there first, which is where you come in.

The field also needs honest practitioners. The history of enterprise risk management is littered with frameworks that looked good in a slide deck and fell apart in implementation, because they were designed by people who hadn't spent enough time in the systems, the audit rooms, and the uncomfortable conversations with CFOs explaining why the controls they believed were working weren't.

Enterprise Risk Orchestration will only develop credibility if it's built on that kind of honesty. The four planes aren't abstract. They came from the field. The gap taxonomy didn't come from research; it came from seeing the same five gaps appear in different organizations for twenty years. The maturity model reflects real progression paths, the ones that work and the ones that stall.

When you apply these frameworks in your organization, you'll find things that don't fit perfectly. That's good. The field develops when practitioners push back on frameworks, extend them, break them, and publish what they learned. The goal isn't compliance with a framework. It's better assurance of whether your enterprise is actually in control.

CONTINUOUS ASSURANCE AS THE NEW NORMAL

The standard is shifting. Not because someone mandated it, though the regulatory direction is clear enough. It's shifting because the cost of the old standard is finally becoming visible.

Periodic assurance was designed for a world where controls changed slowly, transactions were reviewed by humans, and the interval between risk events and detection was measured in weeks. That world is gone. Transactions happen at machine speed. Controls drift without anyone touching them. AI systems make decisions that humans won't see for months, if ever. The annual audit cycle was never designed to provide assurance in that environment, and it doesn't.

The organizations that figure this out first will have a real advantage. Their audit cycles will be faster, because the evidence will already exist and already be organized. Their regulatory exposure will be lower, because they'll find gaps before examiners do. Their board conversations will be different, because they'll be able to answer "are we in control" with evidence rather than attestation.

The CFO who can walk into an audit committee meeting and show a live control confidence score, backed by cross-system evidence, isn't performing assurance theater. They're doing something genuinely new. They're providing the kind of real-time, evidence-based answer that boards have been asking for since Sarbanes-Oxley passed and never quite gotten.

That's the end state. It doesn't happen overnight. The maturity path is real, and most organizations start somewhere in the middle. But the direction is clear, and the first step is always the same: stop accepting fragmentation as inevitable and start treating cross-system coherence as a function.

The tools exist. The data exists. The frameworks are in this book. The only thing that hasn't existed, until now, is a name for the discipline that ties them together.

The category is named.

Enterprise Risk Orchestration is the practice of reading your control environment as a whole, not as a collection of parts. Continuous assurance is the standard that discipline makes possible. Control Intelligence is where it's heading, when the evidence layer becomes smart enough to surface risk before it becomes exposure.

You've spent your career building the pieces. Now you can build the whole.

The conversation starts with the question your board should have been asking all along: can you prove, right now, that your controls are working? If the honest answer is "not really," you know what to do next.

Everybody has been solving a piece. You're now equipped to see the whole.

Chapter 15
Notes and References

This book draws principally on the author's own experience and synthesis across nearly three decades in enterprise risk and controls. The notes below identify the external frameworks, public disclosures, and vendor sources referenced in the text. The specific companies behind the anonymized examples are withheld by design.

The four planes of assurance (front matter)

Sherwood, J., Clark, A., and Lynas, D. Enterprise Security Architecture: A Business-Driven Approach. CMP Books / CRC Press (Taylor and Francis), 2005. ISBN 978-1-57820-318-5.

The SABSA Institute, White Paper W100 (sabsa.org). SABSA is a registered trademark of The SABSA Institute. This book is an independent work and is not affiliated with, endorsed by, or certified by The SABSA Institute; the four-plane model is the author's own synthesis, not a reproduction of the SABSA framework.

Disclosed control failures (A Pile of Tools Is Not a Strategy, and the disclosure table)

The examples are illustrative composites drawn from public U.S. Securities and Exchange Commission Form 10-K, Item 9A ("Controls and Procedures") disclosures of material weaknesses in IT general controls — segregation of duties, user access, and change management — for fiscal years 2019 through 2024. Company identities are withheld; the underlying disclosures are a matter of public record and are searchable through the SEC's EDGAR full-text search system (efts.sec.gov).

Industry data on internal-control failures (Compliant and Exposed, and What the Audit Report Doesn't Say)

Ideagen Audit Analytics. SOX 404 Disclosures: A 19-Year Review, 2004–2022. August 2023. www.auditanalytics.com. An aggregate analysis of internal-control-over-financial-reporting (ICFR) disclosures across SEC registrants, and the source for trend figures on material weaknesses, segregation-of-duties findings, and the rise of information-technology issues to the single most-cited control weakness in adverse auditor attestations for FY2022.

SAP's next-generation GRC direction (A Pile of Tools Is Not a Strategy, and The AI Accelerant)

Verma, V., and Thierer, J. "Announcing SAP's Plans for a Next-Generation Governance, Risk, and Compliance Platform: SAP GRC edition for SAP HANA." SAP Community, GRC Tuesdays blog series, 2023. https://community.sap.com/t5/financial-management-blog-posts-by-sap/grc-tuesdays-announcing-sap-s-plans-for-a-next-generation-governance-risk/ba-p/13570506

"What's New in SAP GRC for SAP HANA." SAP Community, GRC Tuesdays blog series, March 2026. Describes SAP GRC for SAP HANA as a HANA-native unification of SAP GRC 12.0, in restricted shipment from March 2026 with general availability planned for the third quarter of 2026. https://community.sap.com/t5/financial-management-blog-posts-by-sap/grc-tuesdays-what-s-new-in-sap-grc-for-sap-hana/ba-p/14343511

"Latest Trends in Governance, Risk, and Compliance Technology." SAP Community, GRC Tuesdays blog series. https://community.sap.com/t5/financial-management-blog-posts-by-sap/grc-tuesdays-latest-trends-in-governance-risk-and-compliance-technology/ba-p/13570536

SAP's roadmap communications are forward-looking statements of plans that SAP may change at any time without notice; they are referenced here as a directional signal, not a commitment.

Chapter 16
What Comes Next

This book set out to name something: a discipline for keeping the whole enterprise in control, not just its parts. Naming it is the start. Making it run is the work.

That's what I'm building now. Sentralis develops software for Enterprise Risk Orchestration — the discipline this book describes. ControlOS orchestrates fragmented control signals from across your enterprise systems into a single, real-time view of risk. Mitigo is the AI layer that turns that view into answers, and into action, for the people accountable for it.

If the book changed how you see the problem, that's enough. If you want to go further — the field guide, the research, or the work itself — start at mitigosuite.com.

E.H. Schmidt has spent nearly thirty years in governance, risk, and controls across the Big Four, SAP, and Xiting. He is the founder and CEO of Sentralis Inc.

www.ingramcontent.com/pod-product-compliance
Lightning Source LLC
LaVergne TN
LVHW022221200926
848481LV00006B/2057
9798996776306